Springer Optimization and Its Applications

Volume 136

Aims and Scope
Optimization has been expanding in all directions at an astonishing rate during the last few decades. New algorithmic and theoretical techniques have been developed, the diffusion into other disciplines has proceeded at a rapid pace, and our knowledge of all aspects of the field has grown even more profound. At the same time, one of the most striking trends in optimization is the constantly increasing emphasis on the interdisciplinary nature of the field. Optimization has been a basic tool in all areas of applied mathematics, engineering, medicine, economics and other sciences.

The series *Springer Optimization and Its Applications* aims to publish state-of-the-art expository works (monographs, contributed volumes, textbooks) that focus on algorithms for solving optimization problems and also study applications involving such problems. Some of the topics covered include nonlinear optimization (convex and nonconvex), network flow problems, stochastic optimization, optimal control, discrete optimization, multi-objective programming, description of software packages, approximation techniques and heuristic approaches.

More information about this series at http://www.springer.com/series/7393

Jason Papathanasiou • Nikolaos Ploskas

Multiple Criteria Decision Aid

Methods, Examples and Python Implementations

Springer

Jason Papathanasiou
Department of Business Administration
University of Macedonia
Thessaloniki, Greece

Nikolaos Ploskas
Carnegie Mellon University
Pittsburgh, PA, USA

ISSN 1931-6828 ISSN 1931-6836 (electronic)
Springer Optimization and Its Applications
ISBN 978-3-030-06272-9 ISBN 978-3-319-91648-4 (eBook)
https://doi.org/10.1007/978-3-319-91648-4

Mathematics Subject Classification: 90, 90C29, 90C70, 90C90

Printed on acid-free paper

This Springer imprint is published by the registered company Springer Nature Switzerland AG
The registered company address is: Gewerbestrasse 11, 6330 Cham, Switzerland

To Dimitra, Alexander, and Charikleia, always
For their love and patience during sailing in uncharted waters
– Jason Papathanasiou

To my family
– Nikolaos Ploskas

Foreword

I am particularly happy to write this foreword, for several reasons. First, I appreciate the work done by Jason Papathanasiou and Nikolaos Ploskas to propagate multicriteria decision aid in classes, to make students think about the consequences of their decisions, and to promote the use of multicriteria methods in actual decision problems. Inspired by the famous quote by the French writer and politician André Malraux, I am sure that the twenty-first century will be ethical or it will not be. Indeed, ethics is essential for the survival of our world. Decisions shape our future. Decisions are often difficult to make. Decisions can be good or bad. Decisions have often been poorly made, especially during the last decades. Today, we can observe the resulting crises all around the world. Many decisions are bad because they rely on a single, often economic (cost or profit) criterion. They can be made on a qualitative basis (experience, expertise, etc.) or using unicriterion optimization models and methods. Anyway, they usually fail because they are shortsighted and biased: they do not take into account all the stakeholders, all the objectives that are essential for our future. As a crucial example, achieving sustainable development is impossible with a unicriterion approach. It calls for a multicriteria approach encompassing economic as well as social and environmental objectives. Multicriteria decision aid can help us to achieve better, more sustainable decisions. This book is an important step toward a more widespread use of multicriteria models and methods.

A second point is that the authors do not focus on a single method but rather review different multicriteria methods. It is important to understand that there exists no ideal multicriteria decision-making method. Instead many different methods have been proposed over the last 50 years. Each of them has advantages and limits. Each of them is making specific assumptions about the type of the decision problem and about the preferences of the decision makers. Choosing the "right" method also depends on the problem at hand: the availability of data, the quality of data, the type and number of decision makers, the requirements of the different stakeholders, etc. All these parameters have an impact on the choice of an appropriate method. With this in mind, the authors present the principles and the characteristics of six families of methods. This set of methods is not exhaustive; other interesting methods exist as well. However, the choice that they have made reflects the different types of methods currently available as good as possible.

TOPSIS and VIKOR are two compensatory aggregation methods that have been widely used during the recent years. They are simple to use and rather straightforward to understand. PROMETHEE is a well-known outranking method. It allows for partial compensation between criteria and has many extensions including among others advanced sensitivity analysis tools and the GAIA visual descriptive model. It thus provides decision makers with a much richer information at the expense of a more complex preference modeling. The SIR method is another interesting extension of PROMETHEE. On the other hand, AHP is quite different as it is designed to work with qualitative input data and it is based on a very specific pairwise comparison principle. Many people have been critical about the theoretical basis of AHP. Yet, it is one of the most widely used multicriteria decision aid methods. Finally, goal programming methods are based on the mathematical programming model. They are an appropriate choice when decisions are related to decision variables whose best values should be determined under a set of constraints.

The lecture of this book will provide the reader with a quite good overview of existing multicriteria decision-making methods and with guidelines for selecting an appropriate method for each decision problem. It is helpful for the practitioner as well as for the researcher wishing to compare different approaches and methods.

Finally, the authors provide reusable Python computer code for each of the methods presented in the book. The availability of computer solutions is essential for the use of multicriteria methods. When I started working on multicriteria methods back in the 1980s, people had to rely on programming mainframe computers to implement methods. It was difficult, time consuming, and only possible for people with confirmed computer skills. Interaction was very limited and computer graphics were nearly nonexistent. The advent of personal computers started a new era. Around 1990, interactive computer software, such as the PROMCALC implementation of PROMETHEE, appeared. They made it possible for individual users without programming skills to apply multicriteria decision aid methods in actual decision processes. Progressively, more applications appeared and methods evolved based on the feedback from actual applications. At the turn of the century, graphical interfaces became widely available and more advanced, and user-friendly software appeared, boosting even more the use of multicriteria methods. Today, most multicriteria methods are supported by specific software such as Visual PROMETHEE or Expert Choice (for AHP). Practitioners have powerful tools available to perform comprehensive and detailed analysis. However, these specialized pieces of software do not make it easy to perform comparisons between different methods, to adapt methods to special decision problems, or to test new developments or extensions of the methods. The Python code provided by the authors is an elegant and original solution to these shortcomings. It is open source, can be modified easily, and relies on a modern and easy-to-learn language available on multiple platforms. Only moderate computer skills are necessary to take advantage of it. Students will be able to use it to better understand the inner works of the methods and to develop more advanced projects. Practitioners will have a way to adapt the methods to their specific needs. Researchers will have the basis for analyzing the characteristics of the methods, for validating theoretical experiences, or for developing new modules or methods.

In the compact format of this book, the authors manage to transmit essential information needed to learn, understand, and apply multicriteria decision aid. They also provide interested readers with the means to practice, to test, and to validate their acquired knowledge.

Πολλές ευχαριστίες στους Ιάσωνα και Νικόλαο για αυτό το υπέροχο βιβλίο!

(Many thanks to Jason and Nikolaos for this wonderful book!)

Brussels, Belgium
May 2017

Bertrand Mareschal

Preface

The Multiple Criteria Decision Aid (MCDA) discipline is growing rapidly as the number of publications in the literature soared during the recent years. It is successfully applied in all kinds of scientific domains and numerous MCDA methodologies exist with an even larger number of variations available. The rationale behind this book is something that the authors were able to identify at a quite early stage on their engagement in the field of MCDA. And that is that, as already said, there is a lot of research available in the literature; however much of it is difficult to interpret or be exploited by the nonexpert researcher, practitioner, or academic. This is an applied field of research and people need to be able to fully understand each step of the methodologies and implement them easily, coherently, and more importantly correctly.

The main feature of this book is the presentation of a variety of MCDA methods. This book includes the thorough theoretical and computational presentation of six MCDA methods:

- TOPSIS
- VIKOR
- PROMETHEE
- SIR
- AHP
- Goal Programming

The novelty of this book is that the presentation of each method is focused on three aspects:

- Initially, the theoretical background is presented for each method including variants that have been proposed in the literature.
- Secondly, a thorough numerical example is presented for each method.
- Finally, a Python code is presented to fully cover the presentation of each method. The Python implementations that are presented in this book are as simple as possible.

This book is addressed to students, scientists, and practitioners. Students will learn various MCDA methods through illustrative examples, while they will be able to solve the examples using the Python codes given in this book. This book covers thoroughly a course on Multicriteria Decision Analysis whether or not Python is used. Scientists and practitioners will have a book in their library that presents many different MCDA methods and their variants.

The book is organized in six chapters. A simple, yet important method in the MCDA pantheon is presented in the first chapter, namely TOPSIS. The following chapter is about a very similar method called VIKOR, a method that recently has gained much popularity. The algorithms in both cases have a number of distinct steps and various other techniques that can be used to modify the classical versions of the methods, which are also naturally presented. The fuzzy variations are included too, as well as the group decision-making extensions. The third chapter focuses on PROMETHEE, a well-known member of the outranking family of methods. PROMETHEE I and II are fully analyzed and then the text proceeds with PROMETHEE V as well as short references to the GAIA method and the group decision-making PROMETHEE procedure. SIR is next, a method that integrates both TOPSIS and PROMETHEE ideas and could be actually considered as a generalization of PROMETHEE. This is a relative new member of the MCDA portfolio, but the authors feel that SIR could in time make a difference. A book about MCDA cannot be complete if it does not include the AHP method. Despite its somewhat controversial status within the broader MCDA community, AHP has proven to be a very popular method, perhaps the most popular. This is a reason enough for the authors to include a chapter on AHP in this volume including a discussion on the variations and some shortcomings of the method. The final chapter is about goal programming, which is based actually on the linear programming principles; the classical approach is complemented in this book by the weighted, lexicographic, and Chebyshev versions. Appendix includes the revised Simos method for the calculation of the weights associated with the criteria in an MCDA model, a sometimes rather difficult step during the model formulation. Important methodologies (like ELECTRE, MAUT, ANP, and multiple objective programming) are not included and the authors fully acknowledge this shortcoming; if all goes well, this is expected to be rectified in a future, second edition of the present volume.

Various software packages exist for many MCDA methods, both proprietary and free. In all the chapters of this book, code is provided for each and every step of the procedures, and this is also readily available on the book website. The programming language used is Python, a modern language with a numerous and robust user base. The authors assume a basic knowledge of programming principles and especially of the Python language by the reader; however, they have tried to keep the code as simple as possible. Early on, they decided that important capabilities of the language like the object-oriented programming features, decorators or iterators, are not actually important for the purpose of this book. This is a book intended not only for computer science specialists but also for researchers from every field of science that need tools in order to model first and then solve a particular problem that they need to tackle using an MCDA method. Python is well suited for this task;

the authors pondered at the beginning of this endeavor using another fully object-oriented programming language like Java, but this would needlessly complicate the code. The code is kept as simple as possible but is not simplistic; it is divided in modules and the reader is indeed required to have some knowledge of Python and its particular and unique features. Comments are included in the code, as well as commentary on the main text and this combined with the ability of Python to produce very clearly defined and readable code should result in easily read, understood, and applied code listings. Moreover, the readers are very welcome and actually are encouraged to use, modify, and extend the code to better suit their particular demands.

All of the Python packages included in the book are free and open source as is the language itself; this was another prerequisite firmly set by the authors early on. There is no proprietary code anywhere; the idea is that everyone interested should be able to use the code in this book without restrictions. The Python packages included and necessary for some of the code listings in this book are:

- *graphviz*: a library for creating and rendering graph drawings
- *Matplotlib*: a library used for plotting the results of the MCDA methods
- *PuLP*: an open-source Python-based linear programming solver that allows the user to optimize mathematical programming models
- *Pyomo*: an open-source collection of Python software packages for formulating optimization models

The authors would like to thank the publisher's team for their help and patience during this endeavor and Professor Bertrand Mareschal for agreeing to write the foreword to this book. They also hope that the international research community will find the book interesting and of high standards. For any mistakes in the text the authors, of course, acknowledge exclusive and full responsibility. All codes and accompanied material of this book can be downloaded from https://github.com/springer-math/Multiple-Criteria-Decision-Aid/.

Thessaloniki, Greece — Jason Papathanasiou
Pittsburgh, PA, USA — Nikolaos Ploskas
April 2017

Contents

List of Codes

List of Codes

Chapter 1
TOPSIS

1.1 Introduction

TOPSIS is an acronym that stands for 'Technique of Order Preference Similarity to the Ideal Solution' and is a pretty straightforward MCDA method. As the name implies, the method is based on finding an ideal and an anti-ideal solution and comparing the distance of each one of the alternatives to those. It was presented in [7, 14], and can be considered as one of the classical MCDA methods that has received a lot of attention from scholars and researchers. It has been successfully applied in various instances; for a comprehensive state-of-the-art literature review, refer to Behzadian et al. [3]. Table 1.1, adopted from that reference, presents the distribution of papers on TOPSIS by application areas.

The initial methodology was versatile enough to allow for various experiments and modifications; research has focused on the normalization procedure, the proper determination of the ideal and the anti-ideal solution, and the metric used for the calculation of the distances from the ideal and the anti-ideal solution. Studies also emerged using fuzzy [1, 11, 19, 34, 35, 39] or interval numbers [13, 17, 18] and it has been successfully extended to group decision making. This chapter will present the basic methodology with comments on the various possibilities in each step; a discussion on the rank reversal phenomenon will also take place. In all cases, there will be a detailed numerical example and an implementation in Python. Finally, the chapter will conclude with an extension of TOPSIS for group decision making with fuzzy triangular numbers.

Electronic Supplementary Material The online version of this chapter (https://doi.org/10.1007/978-3-319-91648-4_1) contains supplementary material, which is available to authorized users.

J. Papathanasiou, N. Ploskas, *Multiple Criteria Decision Aid*, Springer Optimization and Its Applications 136, https://doi.org/10.1007/978-3-319-91648-4_1

Table 1.1 Distribution of papers on TOPSIS by application areas [3]

Area	N	%
Supply chain management and logistics	74	27.5
Design, engineering, and manufacturing systems	62	23
Business and marketing management	33	12.3
Health, safety, and environment management	28	10.4
Human resources management	24	8.9
Energy management	14	5.2
Chemical engineering	7	2.6
Water resources management	7	2.6
Other topics	20	7.4
Total	269	100

1.2 Methodology

It is originally assumed that a typical decision matrix with m alternatives, $A_1, \ldots, A_m$, and n criteria, $C_1, \ldots, C_n$, is properly formulated first. Initially, each alternative is evaluated with respect to each of the n criteria separately. These evaluations form a decision matrix $X = \left(x_{ij}\right)_{m \times n}$. Let also $W = (w_1, \ldots, w_n)$ be the vector of the criteria weights, where $\sum_{j=1}^{n} w_j = 1$. The TOPSIS algorithm is comprised of the following six steps:

Step 1. Normalization of the Decision Matrix

In order to be able to compare different kinds of criteria the first step is to make them dimensionless, i.e., eliminate the units of the criteria. In the normalized decision matrix, the normalized values of each performance x_{ij} is calculated as

$$r_{ij} = \frac{x_{ij}}{\sqrt{\sum_{i=1}^{m} x_{ij}^2}}, \quad i = 1, \ldots, m, \quad j = 1, \ldots, n \tag{1.1}$$

Apart from this technique (called vector normalization), several alternatives exist; the linear normalization is as

$$r_{ij} = \frac{x_{ij}}{x_j^+}, \quad x_j^+ = \max_j \left(x_{ij}\right), \quad i = 1, 2, \cdots, m, \quad j = 1, 2, \cdots, n \tag{1.2}$$

if the criterion is to be maximized (benefit criteria) and

$$r_{ij} = \frac{x_{ij}}{x_j^-}, \quad x_j^- = \min_j \left(x_{ij}\right), \quad i = 1, 2, \cdots, m, \quad j = 1, 2, \cdots, n \tag{1.3}$$

if the criterion is to be minimized (cost criteria).

Shih et al. [31] present a summary of other variants of the linear normalization and other methods like the non-monotonic normalization. This is an important step than can greatly affect the final ranking; Pavlicic [29] has pondered upon the effects of normalization to the results of MCDA methods. More recently, Jahan and Edwards [16] published a survey on the influence of normalization techniques in ranking, focusing on the material selection process, but their research can be applied in other domains as well. On the other hand, Vafaei et al. [33] analyzed in detail the effects of the normalization process on a case study based on TOPSIS. The avoidance of the rank reversal phenomenon is also focused on this step (to be further discussed later in this chapter).

Step 2. Calculation of the Weighted Normalized Decision Matrix
The weights are the only subjective parameters (as opposed to other MCDA methodologies) taken into account in TOPSIS; the second step is about the multiplication of the normalized decision matrix with the weight associated with each of the criteria.

$$\sum_{j=1}^{n} w_j = 1, \quad j = 1, 2, \cdots, n \tag{1.4}$$

where w_j is the weight of the jth criterion and

$$v_{ij} = w_j r_{ij}, \quad i = 1, \cdots, m, \quad j = 1, \cdots, n \tag{1.5}$$

are the weighted normalized values.

Step 3. Determination of the Ideal (Zenith) and Anti-ideal (Nadir) Solutions
The simplest case is that the ideal and anti-ideal points are fixed by the decision maker, but this should be avoided as it would imply that the decision maker can actually make a credible elicitation of the two points and it would add more subjectivity to the procedure. Another alternative is that the ideal solution (A^*) is

$$A^* = \{v_1^*, v_2^*, \cdots, v_n^*\} = \left\{\left(\max_j v_{ij} | i \in I'\right), \left(\min_j v_{ij} | i \in I''\right)\right\}, \quad i = 1, 2, \cdots, m, \quad j = 1, \cdots, n \tag{1.6}$$

Meaning that the ideal solution comes from collecting the best performances from the normalized decision matrix. Respectively, the anti-ideal solution (A^-) is

$$A^- = \{v_1^-, v_2^-, \cdots, v_n^-\} = \left\{\left(\min_j v_{ij} | i \in I'\right), \left(\max_j v_{ij} | i \in I''\right)\right\}, \quad i = 1, 2, \cdots, m, \quad j = 1, \cdots, n \tag{1.7}$$

In this case and in accordance with the ideal solution, the anti-ideal solution is derived by the worst performances in the normalized decision matrix. I' is associated with benefit criteria and I'' is associated with cost criteria. A third option (others are available in the literature as well) is to use absolute ideal and anti-ideal points, e.g., $A^* = (1, 1, \cdots, 1)$ and $A^- = (0, 0, \cdots, 0)$.

Step 4. Calculation of the Separation Measures
This step is about the calculation of the distances of each alternative from the ideal solution as

$$D_i^* = \sqrt{\sum_{j=1}^{n} \left(v_{ij} - v_j^*\right)^2}, \quad i = 1, 2, \cdots, m, \quad j = 1, 2, \cdots, n \tag{1.8}$$

Similarly, the distances from the anti-ideal solution are calculated as

$$D_i^- = \sqrt{\sum_{j=1}^{n} \left(v_{ij} - v_j^-\right)^2}, \quad i = 1, 2, \cdots, m, \quad j = 1, 2, \cdots, n \tag{1.9}$$

The above is the most popular case and is the classical Euclidean distance. Other metrics have been adopted, e.g., the Hamming Distance [15] and the Manhattan and Chebyshev distances [31].

Step 5. Calculation of the Relative Closeness to the Ideal Solution
The relative closeness C_i^* is always between 0 and 1 and an alternative is best when it is closer to 1. It is calculated for each alternative and is defined as

$$C_i^* = \frac{D_i^-}{D_i^* + D_i^-}, \quad i = 1, 2, \cdots, m \tag{1.10}$$

Step 6. Ranking of the Preference Order
Finally, the alternatives are ranked from best (higher relative closeness value) to worst. The best alternative and the solution to the problem is on the top of the list.

1.2.1 Numerical Example

The facility location (or location-allocation) problem is a well-known and extensively studied problem in the operational research discipline (for comprehensive reviews, see [28, 30]). In this instance, let's assume that a firm is trying to identify the best site out of six possible choices in order to locate a production facility, taking at the same time into account a number of criteria, namely the investment costs, the employment needs, the social impact, and the environmental impact. The first two criteria are quantitative variables with deterministic values, while the other two are qualitative and rated in a typical Likert scale, ranging from 1 (worst) to 7 (best); all the criteria need to be maximized, since we consider the firm to be a public one.

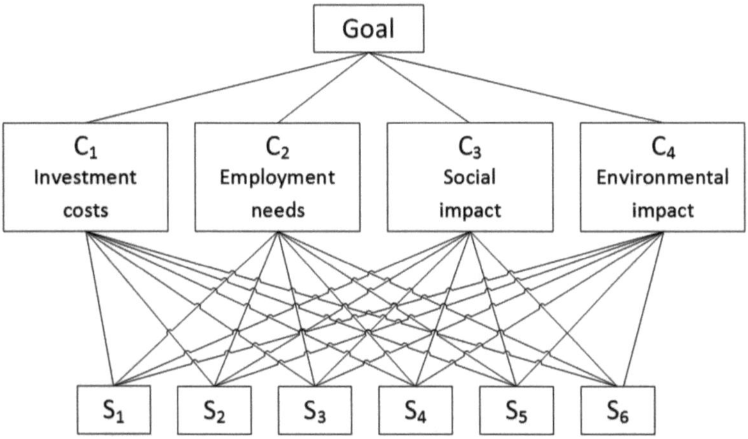

Fig. 1.1 The problem hierarchical structure

Table 1.2 Initial data

	Investment costs (million €)	Employment needs (hundred employees)	Social impact (1–7)	Environmental impact (1–7)
Weight	0.4	0.4	0.1	0.2
Site 1	8	7	2	1
Site 2	5	3	7	5
Site 3	7	5	6	4
Site 4	9	9	7	3
Site 5	11	10	3	7
Site 6	6	9	5	4

If it was private, it would have a different perspective and would most probably be keen on minimizing the first two criteria. The problem structure is shown in Figure 1.1. Table 1.2 includes the weights of each criterion and the performances of the alternatives over all criteria.

Initially, we calculate the normalized decision matrix using the vector normalization technique in Equation (1.1) (Table 1.3). Next, we multiply the normalized decision matrix with the weight associated with each of the criteria using Equation (1.5) (Table 1.4). Then, we determine the ideal and anti-ideal solutions using Equations (1.6) and (1.7) (Table 1.5). Finally, we calculate the distances of each alternative from the ideal and the anti-ideal solution using Equations (1.8) and (1.9) and we also calculate the relative closeness of each alternative using Equation (1.10) (Table 1.6). Figure 1.2 is a graphical representation of Table 1.6. The final ranking of the sites (from best to worst) is Site 5–Site 4–Site 6–Site 3–Site 1–Site 2.

Table 1.3 Normalized decision matrix

	Investment costs	Employment needs	Social impact	Environmental impact
Site 1	0.413	0.377	0.152	0.093
Site 2	0.258	0.162	0.534	0.464
Site 3	0.361	0.269	0.457	0.371
Site 4	0.464	0.485	0.534	0.279
Site 5	0.567	0.538	0.229	0.650
Site 6	0.309	0.485	0.381	0.371

Table 1.4 Weighted normalized decision matrix

	Investment costs	Employment needs	Social impact	Environmental impact
Site 1	0.165	0.113	0.015	0.019
Site 2	0.103	0.048	0.053	0.093
Site 3	0.144	0.081	0.046	0.074
Site 4	0.186	0.145	0.053	0.056
Site 5	0.227	0.162	0.023	0.130
Site 6	0.124	0.145	0.038	0.074

Table 1.5 Ideal and anti-ideal solutions

	Investment costs	Employment needs	Social impact	Environmental impact
Positive ideal solution A_j^*	0.227	0.161	0.053	0.130
Negative ideal solution A_j^-	0.103	0.049	0.015	0.019

Table 1.6 Final ranking for the facility location problem

Site	D_i^*	D_i^-	C_i^*
Site 1	0.141	0.089	0.387
Site 2	0.171	0.083	0.327
Site 3	0.128	0.082	0.389
Site 4	0.086	0.138	0.617
Site 5	0.030	0.201	0.870
Site 6	0.119	0.116	0.493

1.2.2 Python Implementation

The file *TOPSIS_example_1.py* includes a Python implementation that shows how easy it is to use Python in order to obtain the solution of this problem. The input variables are arrays *x* (alternatives performance, lines 9–10) and *weights* (as the name implies, line 13). Comments embedded in the code listing are above each specific step of the TOPSIS procedure. If we omit the blank and comment lines, then we have a solution to this problem in just 22 lines of code; this is just to show how powerful Python is. The normalization method used in lines 18–20 is

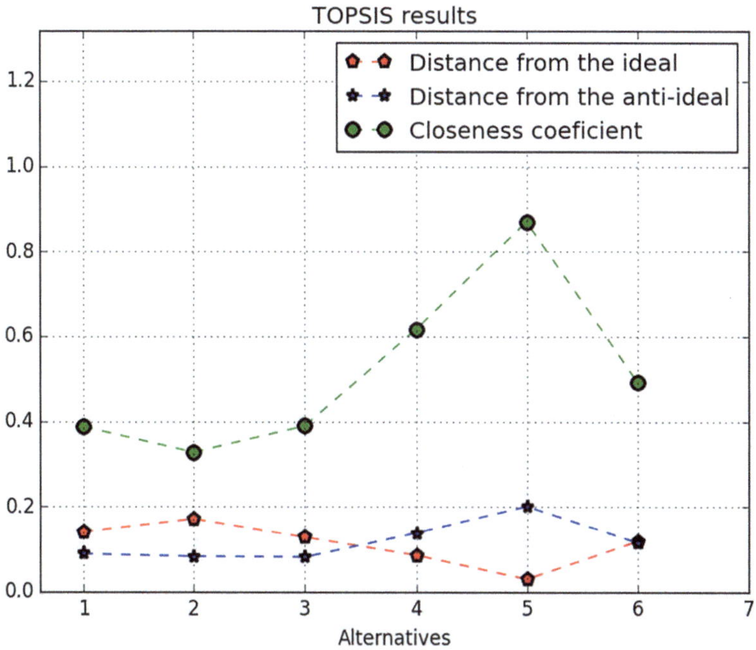

Fig. 1.2 Ideal solution, anti-ideal solution, and closeness coefficient for the facility location problem

the vector normalization and the determination of the ideal and anti-ideal solution is based on finding the best and worst performances (lines 28–31, arrays *pis* and *nis*, respectively) of the normalized matrix. We also use the Euclidean distance to measure the distance of the alternatives to the ideal and anti-ideal solution (lines 35–37 and 41–43, arrays *dpis* and *dnis*, respectively). Finally, the result (relative closeness coefficients) is calculated in lines 47–48 and printed in the console in line 49.

```
1.  # Filename: TOPSIS_example_1.py
2.  # Description: Application of TOPSIS in the first facility
3.  # location problem of Chapter 1
4.  # Authors: Papathanasiou, J. & Ploskas, N.
5.
6.  from numpy import *
7.
8.  # performances of the alternatives
9.  x = array([[8, 7, 2, 1], [5, 3, 7, 5], [7, 5, 6, 4],
10.     [9, 9, 7, 3], [11, 10, 3, 7], [6, 9, 5, 4]])
11.
12. # weights of the criteria
13. weights = array([0.4, 0.3, 0.1, 0.2])
14.
```

```
# Step 1 (vector normalization): cumsum() produces the
# cumulative sum of the values in the array and can also
# be used with a second argument to indicate the axis to use
col_sums = array(cumsum(x**2, 0))
norm_x = array([[round(x[i, j] / sqrt(col_sums[x.shape[0]
    - 1, j]), 3) for j in range(4)] for i in range(6)])

# Step 2 (Multiply each evaluation by the associated weight):
# wnx is the weighted normalized x matrix
wnx = array([[round(weights[i] * norm_x[j, i], 3)
    for i in range(4)] for j in range(6)])

# Step 3 (positive and negative ideal solution)
pis = array([amax(wnx[:, :1]), amax(wnx[:, 1:2]),
    amax(wnx[:, 2:3]), amax(wnx[:, 3:4])])
nis = array([amin(wnx[:, :1]), amin(wnx[:, 1:2]),
    amin(wnx[:, 2:3]), amin(wnx[:, 3:4])])

# Step 4a: determine the distance to the positive ideal
# solution (dpis)
b1 = array([[(wnx[j, i] - pis[i])**2 for i in range(4)]
    for j in range(6)])
dpis = sqrt(sum(b1, 1))

# Step 4b: determine the distance to the negative ideal
# solution (dnis)
b2 = array([[(wnx[j, i] - nis[i])**2 for i in range(4)]
    for j in range(6)])
dnis = sqrt(sum(b2, 1))

# Step 5: calculate the relative closeness to the ideal
# solution
final_solution = array([round(dnis[i] / (dpis[i] + dnis[i]),
    3) for i in range(6)])
print("Closeness coefficient = ", final_solution)
```

Although the code in file *TOPSIS_example_1.py* produces correct results, it is however rather simplistic and does not take full advantage of Python's capabilities. The file *TOPSIS.py* includes a Python implementation that uses the same input and produces the same results, it has however additional functionalities and allows for more experimentations. Apart from *numpy*, it uses the *matplotlib* library to plot the results and the *timeit* module to time the procedure (lines 5–7). Each step is implemented in its own function and the way to use the function, e.g., the input of the function, is embedded as a Python doc string.

More analytically:

- Step 1 (function *norm(x, y)*) is in lines 10–29. It includes two different normalization techniques, namely the vector (lines 15–20) and the linear (lines 21–29) normalization techniques. The inputs are the array with the alternatives performances (variable *x*) and the normalization method (variable *y*). Its output is the normalized decision matrix (variable *z*).

- Step 2 (function *mul_w(r, t)*) is in lines 32–40. The inputs are the criteria weights matrix (variable *r*) and the normalized matrix from Step 1 (variable *t*). Its output is the weighted normalized decision matrix (variable *z*).
- Step 3 (function *zenith_nadir(x, y)*) is in lines 43–61. The inputs are the weighted normalized decision matrix (parameter *x* from Step 2) and the method used to acquire the ideal and anti-ideal points (parameter *y*). For min/max is 'm' (lines 49–57) and anything else for absolute (for clarity, users can enter 'a', lines 58–61). Its outputs are the ideal and anti-ideal solutions for each criterion (variables *b* and *c*, respectively).
- Step 4 (function *distance(x, y, z)*) is in lines 65–76. As inputs, it requires the weighted normalized decision matrix (parameter *x* from Step 2) and the results of the zenith and nadir points from Step 3 (parameters *y* and *z*). Its outputs are the distances from the zenith and nadir points.
- Step 5 (function *topsis(matrix, weight, norm_m, id_sol, pl)*) is in lines 80–110. This function calls all the other functions and produces the final result. Variable *matrix* is the initial decision matrix, *weight* is the criteria weights matrix, *norm_m* is the normalization method choice, *id_sol* is the action used to define the ideal and the anti-ideal solution, and *pl* indicates whether or not to plot the results using *matplotlib* or not. Figure 1.2 was produced using the *matplotlib* library.

Note that the code allows the inclusion of only benefit criteria. It is however easy for interested readers to make the appropriate changes to also allow cost criteria.

```
1.  # Filename: TOPSIS.py
2.  # Description: TOPSIS method
3.  # Authors: Papathanasiou, J. & Ploskas, N.
4.
5.  from numpy import *
6.  import matplotlib.pyplot as plt
7.  import timeit
8.
9.  # Step 1: normalize the decision matrix
10. def norm(x, y):
11.     """ normalization function; x is the array with the
12.     performances and y is the normalization method.
13.     For vector input 'v' and for linear 'l'
14.     """
15.     if y == 'v':
16.         k = array(cumsum(x**2, 0))
17.         z = array([[round(x[i, j] / sqrt(k[x.shape[0] - 1,
18.             j]), 3) for j in range(x.shape[1])]
19.             for i in range(x.shape[0])])
20.         return z
21.     else:
22.         yy = []
23.         for i in range(x.shape[1]):
24.             yy.append(amax(x[:, i:i + 1]))
25.             k = array(yy)
26.         z = array([[round(x[i, j] / k[j], 3)
```

```
            for j in range(x.shape[1])]
            for i in range(x.shape[0])])
        return z

# Step 2: find the weighted normalized decision matrix
def mul_w(r, t):
    """ multiplication of each evaluation by the associate
    weight; r stands for the weights matrix and t for
    the normalized matrix resulting from norm()
    """
    z = array([[round(t[i, j] * r[j], 3)
        for j in range(t.shape[1])]
        for i in range(t.shape[0])])
    return z

# Step 3: calculate the ideal and anti-ideal solutions
def zenith_nadir(x, y):
    """ zenith and nadir virtual action function; x is the
    weighted normalized decision matrix and y is the
    action used. For min/max input 'm' and for absolute
    input enter 'a'
    """
    if y == 'm':
        bb = []
        cc = []
        for i in range(x.shape[1]):
            bb.append(amax(x[:, i:i + 1]))
            b = array(bb)
            cc.append(amin(x[:, i:i + 1]))
            c = array(cc)
        return (b, c)
    else:
        b = ones(x.shape[1])
        c = zeros(x.shape[1])
        return (b, c)

# Step 4: determine the distance to the ideal and anti-ideal
# solutions
def distance(x, y, z):
    """ calculate the distances to the ideal solution (di+)
    and the anti-ideal solution (di-); x is the result
    of mul_w() and y, z the results of zenith_nadir()
    """
    a = array([[(x[i, j] - y[j])**2
       for j in range(x.shape[1])]
       for i in range(x.shape[0])])
    b = array([[(x[i, j] - z[j])**2
       for j in range(x.shape[1])]
        for i in range(x.shape[0])])
    return (sqrt(sum(a, 1)), sqrt(sum(b, 1)))

# TOPSIS method: it calls the other functions and includes
# step 5
def topsis(matrix, weight, norm_m, id_sol, pl):
```

```
    """ matrix is the initial decision matrix, weight is
    the weights matrix, norm_m is the normalization
    method, id_sol is the action used, and pl is 'y'
    for plotting the results or any other string for
    not
    """
    z = mul_w(weight, norm(matrix, norm_m))
    s, f = zenith_nadir(z, id_sol)
    p, n = distance(z, s, f)
    final_s = array([n[i] / (p[i] + n[i])
        for i in range(p.shape[0])])
    if pl == 'y':
        q = [i + 1 for i in range(matrix.shape[0])]
        plt.plot(q, p, 'p--', color = 'red',
            markeredgewidth = 2, markersize = 8)
        plt.plot(q, n, '*--',  color = 'blue',
            markeredgewidth = 2, markersize = 8)
        plt.plot(q, final_s, 'o--', color = 'green',
             markeredgewidth = 2, markersize = 8)
        plt.title('TOPSIS results')
        plt.legend(['Distance from the ideal',
            'Distance from the anti-ideal',
            'Closeness coefficient'])
        plt.xticks(range(matrix.shape[0] + 2))
        plt.axis([0, matrix.shape[0] + 1, 0, 3])
        plt.xlabel('Alternatives')
        plt.legend()
        plt.grid(True)
        plt.show()
    return final_s

# performances of the alternatives
x = array([[8, 7, 2, 1], [5, 3, 7, 5], [7, 5, 6, 4],
           [9, 9, 7, 3], [11, 10, 3, 7], [6, 9, 5, 4]])

# weights of the criteria
w = array([0.4, 0.3, 0.1, 0.2])

# final results
start = timeit.default_timer()
topsis(x, w, 'v', 'm', 'n')
stop = timeit.default_timer()
print("time = ", stop - start)
print("Closeness coefficient = ",
   topsis(x, w, 'v', 'm', 'y'))
```

The results are shown in Table 1.7 and Figure 1.3 if we choose the vector normalization and the calculation of the best and worst performances for the ideal and anti-ideal solutions (case (a), Table 1.7 and Figure 1.3). If we choose the linear normalization and absolute ideal and anti-ideal points, the results are more or less the same for the top three sites (case (b), Table 1.7 and Figure 1.3). However, if we modify the data for the first criterion (investments costs) and change the performance of site 4 from 9 to 1 and of site 5 from 11 to 3, the results are quite

Table 1.7 Solution for the facility location problem using initial data

	Solution using vector normalization and min/max for the ideal and anti-ideal solution (case (a))			Solution using linear normalization and absolute for the ideal and anti-ideal solution (case (b))		
	D_i^*	D_i^-	C_i^*	D_i^*	D_i^-	C_i^*
Site 1	0.141	0.089	0.387	1.736	0.361	0.172
Site 2	0.171	0.083	0.327	1.744	0.268	0.133
Site 3	0.128	0.082	0.389	1.703	0.328	0.161
Site 4	0.086	0.138	0.617	1.622	0.444	0.215
Site 5	0.030	0.201	0.870	1.551	0.540	0.258
Site 6	0.119	0.116	0.493	1.671	0.372	0.182

Table 1.8 Solution for the facility location problem using modified data

	Solution using vector normalization and min/max for the ideal and anti-ideal solution (case (c))			Solution using linear normalization and absolute for the ideal and anti-ideal solution (case (d))		
	D_i^*	D_i^-	C_i^*	D_i^*	D_i^-	C_i^*
Site 1	0.127	0.216	0.630	1.694	0.454	0.211
Site 2	0.147	0.144	0.495	1.713	0.318	0.157
Site 3	0.102	0.190	0.650	1.663	0.407	0.197
Site 4	0.219	0.111	0.335	1.755	0.305	0.148
Site 5	0.151	0.168	0.527	1.664	0.393	0.191
Site 6	0.084	0.186	0.689	1.634	0.425	0.207

different and the ranking changes considerably between the versions of TOPSIS and for the same data set, as seen in Table 1.8 and Figure 1.3 (cases (c) and (d)). The reader is encouraged to experiment further with the various variants of TOPSIS and compare the results. An interesting note can be made on the running time of the TOPSIS algorithm; using the *timeit* module in lines 120–122 and calling function *topsis* without printing the results (the value of variable *pl* is set to *'n'*), the running time is (an average of ten runs) 0.00085 s on a Linux machine with an Intel Core i7 at 2.2 GHz CPU and 6 GB RAM. Running the same code with 1,000 sites and four criteria, the execution time is 0.1029 s.

1.2.3 Rank Reversal

TOPSIS is also not immune to a phenomenon called rank reversal. This is a phenomenon especially studied for other MCDA methods and will be mentioned again during the course of this book. It has produced much debate among the scientific community and continues to be considered controversial by many. In its simplest form, it occurs when an alternative is added to the initial list of alternatives and then the final ranking changes considerably from the original one; in some cases,

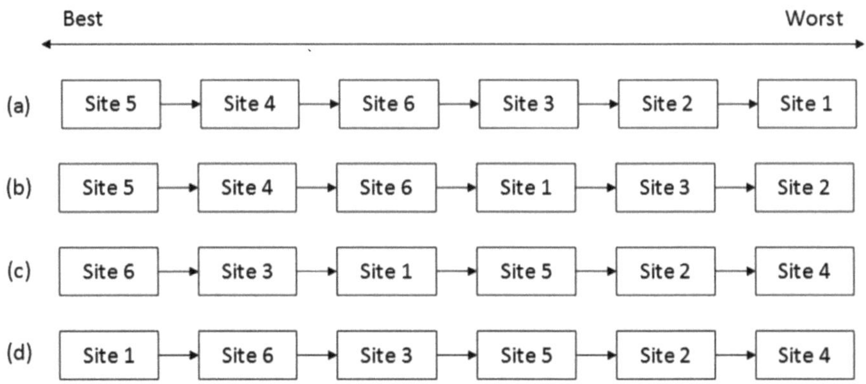

Fig. 1.3 Solutions for the facility location problem cases (a) and (b) with initial data and cases (c) and (d) with modified data

Table 1.9 Initial ranking in the rank reversal example

	$w_1 = 0.5$	$w_2 = 0.5$				
	C_1	C_2	D_i^*	D_i^-	C_i^*	Ranking
A_1	1	5	0.294	0.244	0.453	3
A_2	4	2	0.244	0.294	0.546	1
A_3	3	3	0.189	0.212	0.529	2

we observe that the final ranking is totally reversed. Wang and Luo [37] studied it for a number of methods, including TOPSIS, and concluded by saying that it might be, after all, a normal phenomenon. Yet, they do not provide a solution, while Garcia-Cascales and Lamata [12] propose modifications in the original algorithm in order to avoid it. In the next part of this section, we will recreate the example used by Garcia-Cascales and Lamata [12] to demonstrate the phenomenon.

They consider three different candidates for a certain position, individuals A_1, A_2, and A_3. Each one of them completes two questionnaires with the same weight ($w_1 = 0.5$ and $w_2 = 0.5$). Therefore, the formulated multiple criteria problem has a couple of criteria and three alternatives. The initial input data and the final ranking after applying TOPSIS (using the vector normalization and the calculation of the best and worst performances for the ideal and anti-ideal solutions) are shown in Table 1.9; the ranking is $A_2 > A_3 > A_1$.

Let's add now another candidate for the position, A_4, whose responses to the questionnaires are evaluated to 5 and 1 for C_1 and C_2, respectively (Table 1.10). We observe that the ranking has changed; it is now $A_1 > A_3 > A_2 > A_4$. A_1, which was originally the worst, is now the best; the order is totally reversed! This does not seem logical in many decision problems and is the root of many heated debates.

As already mentioned, the normalization used in their example is the vector normalization, while they determine the ideal and anti-ideal solutions using the best and worst performances. To avoid this phenomenon, they apply the linear

Table 1.10 Ranking after the introduction of a new alternative in the rank reversal example

	$w_1 = 0.5$	$w_2 = 0.5$				
	C_1	C_2	D_i^*	D_i^-	C_i^*	Ranking
A_1	1	5	0.280	0.320	0.533	1
A_2	4	2	0.250	0.225	0.473	3
A_3	3	3	0.213	0.213	0.500	2
A_4	5	1	0.320	0.280	0.467	4

normalization and modify the definitions of the ideal and anti-ideal solutions with the introduction of two fictitious alternatives. From the example, if we refer to the set $S = [1, 2, 3, 4, 5]$, the values of A^* and A^- would be $(5, 5)$ and $(0, 0)$, respectively; these are the new fictitious alternatives and the rank reversal phenomenon can be avoided (until proven otherwise).

1.3 Fuzzy TOPSIS for Group Decision Making

1.3.1 Preliminaries: Elements of Fuzzy Numbers Theory

In this subsection, we will briefly review important elements from the fuzzy numbers theory, adopted mainly from [2, 23, 40, 41]. The aim of this section is not to make even a short introduction to fuzzy theory; just to present a few elements enough for the reader to keep up with the rest of this chapter. A fuzzy set is a class with a continuum of membership grades [40]; so, a fuzzy set A in a referential (universe of discourse) X is characterized by a membership function A, which associates with each element $x \in X$ a real number $A(x) \in [0, 1]$, having the interpretation $A(x)$ as the membership grade of x in the fuzzy set A.

Let's consider now a fuzzy subset of the real line u: $R \rightarrow [0, 1]$. u is a fuzzy number [2, 10], if it satisfies the following properties:

- u is normal, i.e., $\exists x_0 \in R$ with $u(x_0) = 1$.
- u is fuzzy convex, i.e., $u(tx+(1-t)y) \geq \min\{u(x), u(y)\}, \forall t \in [0, 1], x, y \in R$.
- u is upper semi-continuous on R, i.e., $\forall \epsilon > 0, \exists \delta > 0$ such that $u(x) - u(x_0) < \epsilon, |x - x_0| < \delta$.
- u is compactly supported, i.e., $cl\{x \in R; u(x) > 0\}$ is compact, where $cl(A)$ denotes the closure of the set A.

One of the most popular shapes of fuzzy numbers is the triangular fuzzy number that can be defined as a triplet $A = (\alpha_1, \alpha_2, \alpha_3)$, shown in Figure 1.4. The membership functions are as follows:

Fig. 1.4 Triangular fuzzy number $A = (\alpha_1, \alpha_2, \alpha_3)$, adopted from [23]

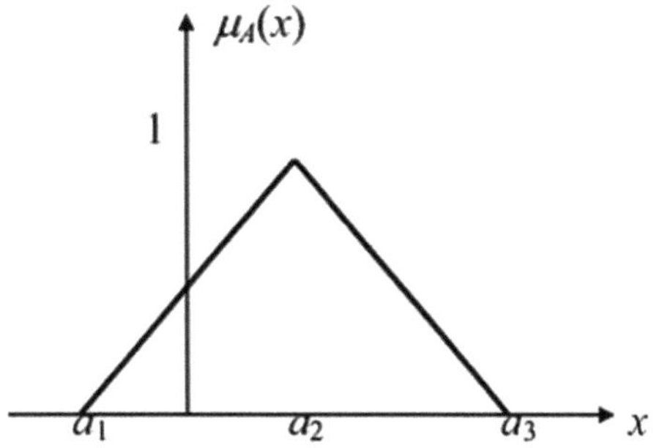

$$\mu_A(x) = \begin{cases} 0, & x < \alpha_1 \\ \frac{x-\alpha_1}{\alpha_2-\alpha_1}, & \alpha_1 \le x \le \alpha_2 \\ \frac{\alpha_3-x}{\alpha_3-\alpha_2}, & \alpha_2 \le x \le \alpha_3 \\ 0, & x > \alpha_3 \end{cases} \tag{1.11}$$

Assuming that both $A = (\alpha_1, \alpha_2, \alpha_3)$ and $B = (b_1, b_2, b_3)$ are triangular fuzzy numbers, then:

- the result from their addition or subtraction is also a triangular fuzzy number.
- the result from their multiplication or division is not a triangular fuzzy number (however, we can assume that the result of multiplication or division is a triangular fuzzy number as an approximation value).
- the max or min operation does not result to a triangular fuzzy number.

Focusing on the addition and the subtraction, they are as

$$A(+)B = (\alpha_1, \alpha_2, \alpha_3)(+)(b_1, b_2, b_3) = (\alpha_1 + b_1, \alpha_2 + b_2, \alpha_3 + b_3) \tag{1.12}$$

and

$$A(-)B = (\alpha_1, \alpha_2, \alpha_3)(-)(b_1, b_2, b_3) = (\alpha_1 - b_3, \alpha_2 - b_2, \alpha_3 - b_1) \tag{1.13}$$

A fuzzy vector is a certain vector that includes an element and has a value between 0 and 1. Bearing this in mind, a fuzzy matrix is a gathering of such vectors. The operations on given fuzzy matrices $A = (\alpha_{ij})$ and $B = (b_{ij})$ are

- sum

$$A + B = \max\left[\alpha_{ij}, b_{ij}\right] \tag{1.14}$$

- max product

$$A \cdot B = \max_k\left[\min\left(\alpha_{ik}, b_{kj}\right)\right] \tag{1.15}$$

- scalar product

$$\lambda A \tag{1.16}$$

where $0 \leq \lambda \leq 1$.

Chen [5] uses the vertex method to calculate the distance between two triangular fuzzy numbers $\widetilde{m} = (m_1, m_2, m_3)$ and $\widetilde{n} = (n_1, n_2, n_3)$ as

$$d(\widetilde{m}, \widetilde{n}) = \sqrt{\frac{1}{3}\left[(m_1 - n_1)^2 + (m_2 - n_2)^2 + (m_3 - n_3)^2\right]} \tag{1.17}$$

Finally, according to Zadeh [41], a linguistic variable is one whose values are words or sentences in a natural or artificial language; among others, he provides an example in the form of the linguistic variable 'Age' that can take the values young, not young, very young, quite young, old, not very old and not very young, etc., rather than 20, 21, 22, 23, $\cdots$. This kind of variables can well be represented by triangular fuzzy numbers. Lee [23] further denotes that a linguistic variable can be defined by the quintuple

$$\text{Linguistic variable} = (x, T(x), U, G, M) \tag{1.18}$$

where:

- x: name of variable
- $T(x)$: set of linguistic terms that can be a value of the variable
- U: set of universe of discourse, which defines the characteristics of the variable
- G: syntactic grammar that produces terms in $T(x)$
- M: semantic rules, which map terms in $T(x)$ to fuzzy sets in U

1.3.2 Fuzzy TOPSIS Methodology

This section presents a fuzzy extension of TOPSIS that is based on Chen's methodology [5], despite the fact that his work has received some criticism by Mahdavi [27]. Variations of fuzzy TOPSIS focus on the distance measurement, the determination of the ideal and anti-ideal points, and the use of other than triangular fuzzy numbers, like trapezoidal fuzzy numbers [24]. However, we use in this section Chen's work due to its relative simplicity as regards to the distance measures compared to the other fuzzy TOPSIS approaches. In addition to the fuzzy numbers (and to complicate things a little bit), TOPSIS is also extended into group decision making involving the opinions of a number of independent experts. Table 1.11, adopted from [20], presents a comparison of the various fuzzy TOPSIS methods proposed in the literature.

Table 1.11 A comparison of fuzzy TOPSIS methods [20]

Source	Attribute weights	Type of fuzzy numbers	Ranking method	Normalization method
Chen and Hwang [7]	Fuzzy numbers	Trapezoidal	Lee and Li's [22] generalized mean method	Linear normalization
Liang [25]	Fuzzy numbers	Trapezoidal	Chen's [6] ranking with maximizing set and minimizing set	Linear normalization
Chen [5]	Fuzzy numbers	Triangular	Chen [5] proposes the vertex method	Linear normalization
Chu [8]	Fuzzy numbers	Triangular	Liou and Wang's [26] ranking method of total integral value with $\alpha = 1/2$	Modified manhattan distance
Tsaur et al. [32]	Crisp values	Triangular	Zhao and Govind's [43] center of area method	Vector normalization
Zhang and Lu [42]	Crisp values	Triangular	Chen's [5] vertex mode	Manhattan distance
Chu and Lin [9]	Fuzzy numbers	Triangular	Kaufmann and Gupta's [21] mean of the removals method	Linear normalization
Cha and Yung [4]	Crisp values	Triangular	Cha and Young [4] propose a fuzzy distance operator	Linear normalization
Yang and Hung [38]	Fuzzy numbers	Triangular	Chen's [5] vertex method	Normalized fuzzy linguistic ratings are used
Wang and Elhag [36]	Fuzzy numbers	Triangular	Chen's [5] vertex method	Linear normalization
Jahanshahloo et al. [18]	Crisp values	Interval data	Jahanshahloo et al. [18] propose a new column and ranking method	

In conjunction with the steps of the typical TOPSIS method presented earlier in this chapter, the steps of the fuzzy extension are:

Step 1. Form a Committee of Decision Makers, Then Identify the Evaluation Criteria

If we assume that the decision group has K persons, then the importance of the criteria and the ratings of the alternatives can be calculated as

$$\widetilde{x}_{ij} = \frac{1}{K}\left[\widetilde{x}_{ij}^{1}(+)\widetilde{x}_{ij}^{2}(+)\cdots(+)\widetilde{x}_{ij}^{K}\right] \tag{1.19}$$

$$\widetilde{w}_{j} = \frac{1}{K}\left[\widetilde{w}_{j}^{1}(+)\widetilde{w}_{j}^{2}(+)\cdots(+)\widetilde{w}_{j}^{K}\right] \tag{1.20}$$

where $\widetilde{x}_{ij}^{K}$ and $\widetilde{w}_{j}^{K}$ are the ratings and criteria weights of the Kth decision maker.

Step 2. Choose the Linguistic Variables
Choose the appropriate linguistic variables for the importance weight of the criteria and the linguistic ratings for alternatives with respect to the criteria.

Step 3. Perform Aggregations
Aggregate the weight of criteria to get the aggregated fuzzy weight $\widetilde{w}_j$ of criterion C_j, and pool the decision makers' opinions to get the aggregated fuzzy rating $\widetilde{x}_{ij}$ of alternative A_i under criterion C_j.

Step 4. Construct the Fuzzy Decision Matrix and the Normalized Fuzzy Decision Matrix
The fuzzy decision matrix is constructed as

$$\widetilde{D} = \begin{bmatrix} \widetilde{x}_{11} & \widetilde{x}_{12} & \cdots & \widetilde{x}_{1n} \\ \widetilde{x}_{21} & \widetilde{x}_{22} & \cdots & \widetilde{x}_{2n} \\ \vdots & \vdots & \ddots & \vdots \\ \widetilde{x}_{m1} & \widetilde{x}_{m2} & \cdots & \widetilde{x}_{mn} \end{bmatrix} \tag{1.21}$$

and the vector of the criteria weights as

$$\widetilde{W} = [\widetilde{w}_1, \widetilde{w}_2, \cdots, \widetilde{w}_n] \tag{1.22}$$

where $\widetilde{x}_{ij}$ and $\widetilde{w}_j, i = 1, 2, \cdots, m, j = 1, 2, \cdots, n$ are linguistic variables according to Step 2. They can be described by the triangular fuzzy numbers $\widetilde{x}_{ij} = (a_{ij}, b_{ij}, c_{ij})$ and $\widetilde{w}_j = (w_{j1}, w_{j2}, w_{j3})$. For the normalization step, Chen uses the linear scale transformation in order to drop the units and make the criteria comparable; it is also important to preserve the property stating that the ranges of the normalized triangular fuzzy numbers belong to [0, 1]. The normalized fuzzy decision matrix denoted by $\widetilde{R}$ is

$$\widetilde{R} = [\widetilde{r}_{ij}]_{m \times n} \tag{1.23}$$

The set of benefit criteria is B and the set of cost criteria is C, therefore

$$\widetilde{r}_{ij} = \left(\frac{a_{ij}}{c_j^*}, \frac{b_{ij}}{c_j^*}, \frac{c_{ij}}{c_j^*} \right), \quad j \in B, \quad i = 1, 2, \cdots, m \tag{1.24}$$

$$\widetilde{r}_{ij} = \left(\frac{a_j^-}{c_{ij}}, \frac{a_j^-}{b_{ij}}, \frac{a_j^-}{a_{ij}} \right), \quad j \in C, \quad i = 1, 2, \cdots, m \tag{1.25}$$

$$c_j^* = \max_i c_{ij}, \text{ if } j \in B, \quad i = 1, 2, \cdots, m \tag{1.26}$$

$$a_j^- = \min_i a_{ij}, \text{ if } j \in C, \quad i = 1, 2, \cdots, m \tag{1.27}$$

Step 5. Construct the Fuzzy Weighted Normalized Decision Matrix
Then, the fuzzy weighted normalized decision matrix can be constructed as

$$\widetilde{V} = \left[\widetilde{v}_{ij}\right]_{m\times n}, \quad i = 1, 2, \cdots, m, \quad j = 1, 2, \cdots, n \tag{1.28}$$

where

$$\widetilde{v}_{ij} = \widetilde{r}_{ij}(\cdot)\widetilde{w}_j, \quad i = 1, 2, \cdots, m, \quad j = 1, 2, \cdots, n \tag{1.29}$$

The elements $\widetilde{v}_{ij}, i = 1, 2, \cdots, m, j = 1, 2, \cdots, n$, are normalized positive triangular fuzzy numbers ranging from 0 to 1.

Step 6. Determine the Fuzzy Positive Ideal Solution (FPIS) and the Fuzzy Negative Ideal Solution (FNIS)
The fuzzy positive ideal solution (FPIS, A^*) and the fuzzy negative ideal solution (FNIS, A^-) are

$$A^* = \left(\widetilde{v}_1^*, \widetilde{v}_2^*, \cdots, \widetilde{v}_n^*\right) \tag{1.30}$$

$$A^- = \left(\widetilde{v}_1^-, \widetilde{v}_2^-, \cdots, \widetilde{v}_n^-\right) \tag{1.31}$$

where

$$\widetilde{v}_j^* = (1, 1, 1), \widetilde{v}_j^- = (0, 0, 0), \quad j = 1, 2, \cdots, n \tag{1.32}$$

Step 7. Calculate the Distance of Each Alternative from FPIS and FNIS
The distance of each of the alternatives from FPIS and FNIS can be calculated as

$$D_i^* = \sum_{j=1}^{n} d\left(\widetilde{v}_{ij}, \widetilde{v}_j^*\right), \quad i = 1, 2, \cdots, m, \quad j = 1, 2, \cdots, n \tag{1.33}$$

$$D_i^- = \sum_{j=1}^{n} d\left(\widetilde{v}_{ij}, \widetilde{v}_j^-\right), \quad i = 1, 2, \cdots, m, \quad j = 1, 2, \cdots, n \tag{1.34}$$

where d is the distance measurement between two fuzzy numbers (Equation (1.17)).

Step 8. Calculate the Closeness Coefficient of Each Alternative
The closeness coefficient of each alternative can be defined as

$$CC_i = \frac{D_i^-}{D_i^* + D_i^-}, \quad i = 1, 2, \cdots, m \tag{1.35}$$

Step 9. Rank the Alternatives
An alternative A_i is better than A_j if its closeness coefficient is closer to 1.

Therefore, the final ranking of the alternatives is defined by the value of the closeness coefficient.

1.3.3 Numerical Example

The problem in hand is the same as in the previous examples, the facility location problem; again there are four criteria with six alternative sites (Figure 1.1), but there are now three individual decision makers with different views to be taken into account. In this case, the importance weights of the qualitative criteria and the ratings are considered as linguistic variables expressed in positive triangular fuzzy numbers, as shown in Tables 1.12 and 1.13; they are also considered to be evaluated by decision makers that are experts on the field. These evaluations are in Tables 1.14 and 1.15.

Initially, we aggregate the weights of criteria to get the aggregated fuzzy weights and the ratings to calculate the fuzzy decision matrix using Equations (1.19) and (1.20) (Table 1.16). Next, we calculate the fuzzy normalized decision matrix using Equations (1.24)–(1.27) (Table 1.17). Then, we construct the fuzzy weighted

Table 1.12 Linguistic variables for the criteria

Linguistic variables for the importance weight of each criterion	
Very low (VL)	(0, 0, 0.1)
Low (L)	(0, 0.1, 0.3)
Medium low (ML)	(0.1, 0.3, 0.5)
Medium (M)	(0.3, 0.5, 0.7)
Medium high (MH)	(0.5, 0.7, 0.9)
High (H)	(0.7, 0.9, 1.0)
Very high (VH)	(0.9, 1.0, 1.0)

Table 1.13 Linguistic variables for the ratings

Linguistic variables for the ratings	
Very poor (VP)	(0, 0, 1)
Poor (P)	(0, 1, 3)
Medium poor (MP)	(1, 3, 5)
Fair (F)	(3, 5, 7)
Medium good (MG)	(5, 7, 9)
Good (G)	(7, 9, 10)
Very good (VG)	(9, 10, 10)

Table 1.14 The importance weight of the criteria for each decision maker

	D_1	D_2	D_3
Investment costs	H	VH	VH
Employment needs	M	H	VH
Social impact	M	MH	ML
Environmental impact	H	VH	MH

Table 1.15 The ratings of the six sites by the three decision makers for the four criteria

Criteria	Candidate sites	Decision makers		
		D_1	D_2	D_3
Investment costs	Site 1	VG	G	MG
	Site 2	MP	F	F
	Site 3	MG	MP	F
	Site 4	MG	VG	VG
	Site 5	VP	P	G
	Site 6	F	G	G
Employment needs	Site 1	F	MG	MG
	Site 2	F	VG	G
	Site 3	MG	MG	VG
	Site 4	G	G	VG
	Site 5	P	VP	MP
	Site 6	F	MP	MG
Social impact	Site 1	P	P	MP
	Site 2	MG	VG	G
	Site 3	MP	F	F
	Site 4	MG	VG	G
	Site 5	G	G	VG
	Site 6	VG	MG	F
Environmental impact	Site 1	G	VG	G
	Site 2	MG	F	MP
	Site 3	MP	P	P
	Site 4	VP	F	P
	Site 5	G	MG	MG
	Site 6	P	MP	F

Table 1.16 Fuzzy decision matrix and fuzzy weights

	Investment costs	Employment needs	Social impact	Environmental impact
Site 1	(7.000, 8.667, 9.667)	(4.333, 6.333, 8.333)	(0.333, 1.667, 3.667)	(7.667, 9.333, 10.000)
Site 2	(2.333, 4.333, 6.333)	(6.333, 8.000, 9.000)	(7.000, 8.667, 9.667)	(3.000, 5.000, 7.000)
Site 3	(3.000, 5.000, 7.000)	(6.333, 8.000, 9.333)	(2.333, 4.333, 6.333)	(0.333, 1.667, 3.667)
Site 4	(7.667, 9.000, 9.667)	(7.667, 9.333, 10.000)	(7.000, 8.667, 9.667)	(1.000, 2.000, 3.667)
Site 5	(2.333, 3.333, 4.667)	(0.333, 1.333, 3.000)	(7.667, 9.333, 10.000)	(5.667, 7.667, 9.333)
Site 6	(5.667, 7.667, 9.000)	(3.000, 5.000, 7.000)	(5.667, 7.333, 8.667)	(1.333, 3.000, 5.000)
Weight	(0.833, 0.967, 1.000)	(0.633, 0.800, 0.900)	(0.300, 0.500, 0.700)	(0.700, 0.867, 0.967)

normalized decision matrix using Equations (1.28) and (1.29) (Table 1.18). Finally, we calculate the distance of each alternative from the fuzzy positive ideal and anti-ideal solutions, and calculate the closeness coefficient of each alternative (Table 1.19). Figure 1.5 is a graphical representation of Table 1.19. The final ranking of the sites (from best to worst) is Site 1–Site 4–Site 2–Site 6–Site 5–Site 3.

Table 1.17 Fuzzy normalized decision matrix

	Investment costs	Employment needs	Social impact	Environmental impact
Site 1	(0.700, 0.867, 0.967)	(0.433, 0.633, 0.833)	(0.033, 0.167, 0.367)	(0.767, 0.933, 1.000)
Site 2	(0.233, 0.433, 0.633)	(0.633, 0.800, 0.900)	(0.700, 0.867, 0.967)	(0.300, 0.500, 0.700)
Site 3	(0.300, 0.500, 0.700)	(0.633, 0.800, 0.933)	(0.233, 0.433, 0.633)	(0.033, 0.167, 0.367)
Site 4	(0.767, 0.900, 0.967)	(0.767, 0.933, 1.000)	(0.700, 0.867, 0.967)	(0.100, 0.200, 0.367)
Site 5	(0.233, 0.333, 0.467)	(0.033, 0.133, 0.300)	(0.767, 0.933, 1.000)	(0.567, 0.767, 0.933)
Site 6	(0.567, 0.767, 0.900)	(0.300, 0.500, 0.700)	(0.567, 0.733, 0.867)	(0.133, 0.300, 0.500)

Table 1.18 Fuzzy weighted normalized decision matrix

	Investment costs	Employment needs	Social impact	Environmental impact
Site 1	(0.700, 0.867, 0.967)	(0.433, 0.633, 0.833)	(0.033, 0.167, 0.367)	(0.767, 0.933, 1.000)
Site 2	(0.233, 0.433, 0.633)	(0.633, 0.800, 0.900)	(0.700, 0.867, 0.967)	(0.300, 0.500, 0.700)
Site 3	(0.300, 0.500, 0.700)	(0.633, 0.800, 0.933)	(0.233, 0.433, 0.633)	(0.033, 0.167, 0.367)
Site 4	(0.767, 0.900, 0.967)	(0.767, 0.933, 1.000)	(0.700, 0.867, 0.967)	(0.100, 0.200, 0.367)
Site 5	(0.233, 0.333, 0.467)	(0.033, 0.133, 0.300)	(0.767, 0.933, 1.000)	(0.567, 0.767, 0.933)
Site 6	(0.567, 0.767, 0.900)	(0.300, 0.500, 0.700)	(0.567, 0.733, 0.867)	(0.133, 0.300, 0.500)

Table 1.19 The final result

Alternative	D_i^*	D_i^-	Closeness coefficient	Final ranking
Site 1	1.965	2.305	0.540	1
Site 2	2.212	2.051	0.481	3
Site 3	2.577	1.673	0.394	6
Site 4	1.959	2.279	0.538	2
Site 5	2.526	1.704	0.403	5
Site 6	2.347	1.914	0.449	4

1.3.4 Python Implementation

The file *FuzzyTOPSIS.py* includes a Python implementation of the fuzzy TOPSIS method. Similar to the implementation of TOPSIS, it uses the *matplotlib* library to plot the results and the *timeit* module to time the procedure (lines 6–7). Each step is implemented in its own function and the way to use the functions, e.g., the input of the function, is embedded as a Python doc string.

More analytically:

- Steps 1 and 2 of the fuzzy TOPSIS procedure are in lines 168–198. Dictionaries *cw* (lines 168–171) and *r* (lines 175–177) correspond to Tables 1.12 and 1.13, respectively; they are the definitions of the linguistic variables used for the criteria weights and the ratings. Python list *cdw* (lines 180–181) is about the importance weight of the criteria (Table 1.14) and the ratings of the six candidate sites by the three decision makers are each in its own list, named *c1*, *c2*, $\cdots$, *c6* (lines 185–196, Table 1.15). These lists are concatenated into one list in line 198.
- Steps 3 and 4 are in lines 119–122. Arrays *fuzzy_weights* and *fuzzy_decision_matrix* are the output of function *cal(a, b, k)*. Variable *a* is the dictionary with the

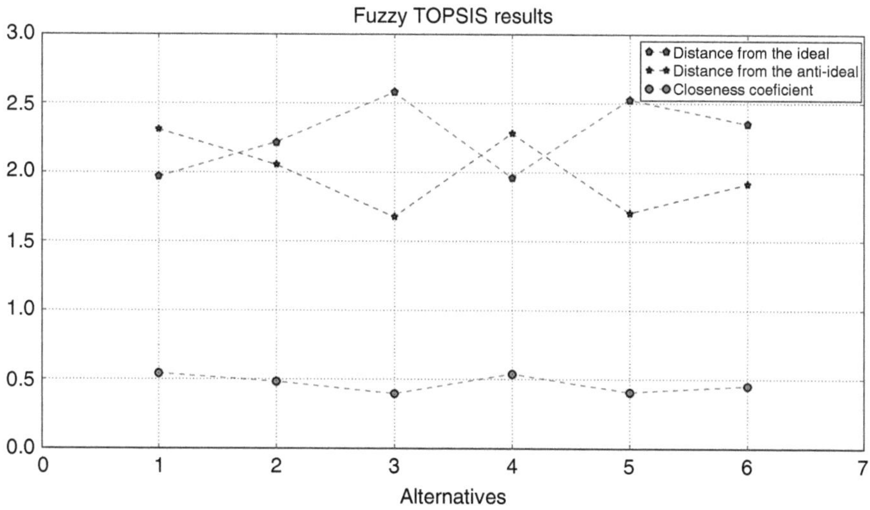

Fig. 1.5 The final result

linguistic variables for the criteria weights or the linguistic variables for the ratings, variable b is the matrix with the criteria weights or the ratings, and variable k is the number of the decision makers. The output is the fuzzy weights of the criteria or the fuzzy decision matrix (Table 1.16). Array *fuzzy_norm_decision_matrix* is the output of function *fndm(a, n, m)*, which uses as an input parameter the output of function *cal*. Variable a is the fuzzy decision matrix, variable n is the number of criteria, and variable m is the number of the alternatives. The output of function *fndm* is the fuzzy normalized decision matrix (Table 1.17).

- Step 5 is in lines 125–127. The fuzzy weighted normalized decision matrix is calculated by calling function *weighted_fndm(a, b, n, m)* that uses as input the results of the previous steps (Table 1.18). Variable a is the fuzzy normalized decision matrix, variable b is the criteria weights, variable n is the number of criteria, and variable m is the number of the alternatives. The output is the fuzzy weighted normalized decision matrix.
- Steps 6 and 7 are in lines 130–133. They make use of functions *func_dist_fpis(a, n, m)* and *func_dist_fnis(a, n, m)*. The first function is about finding the distance from the ideal solution and the second finding the distance from the anti-ideal solution (Table 1.19). Variable a is the fuzzy weighted normalized decision matrix, variable n is the number of criteria, and variable m is the number of the alternatives. The output is the ideal or anti-ideal solution of each criterion. Both functions use function *distance(a, b)* to calculate the distance between two fuzzy triangular numbers. Variables a and b are fuzzy triangular numbers and the output is their distance.
- Step 8 is in lines 136–139 and is about finding the final solution (Table 1.19 and Figure 1.5).

- Final results (function *f_topsis(a, b, c, d, n, m, k, pl)*) are calculated in lines 107–159. This function calls all the other functions and produces the final result. Variable *a* is the dictionary with the linguistic variables for the criteria weights, variable *b* is the matrix with the importance weights of the criteria, variable *c* is a dictionary with the linguistic variables for the ratings, variable *d* is the matrix with all the ratings, variable *n* is the number of criteria, variable *m* is the number of the alternatives, variable *k* is the number of the decision makers, and variable *pl* is whether to plot the results using *matplotlib* or not. Figure 1.5 was produced using the *matplotlib* library.

Similarly to the implementation of TOPSIS, the code of fuzzy TOPSIS allows the inclusion of only benefit criteria. It is however easy for interested readers to make the appropriate changes to also allow cost criteria.

Using the *timeit* module in lines 201–204 and calling function *f_topsis* without printing the results (the value of variable *pl* is set to *'n'*), the running time is (an average of ten runs) 0.0024 s on a Linux machine with an Intel Core i7 at 2.2 GHz CPU and 6 GB RAM. Running the same code with 1,000 sites, four criteria, and three decision makers, the execution time is 0.4015 s.

```
1.   # Filename: FuzzyTOPSIS.py
2.   # Description: Fuzzy TOPSIS method
3.   # Authors: Papathanasiou, J. & Ploskas, N.
4.
5.   from numpy import *
6.   import matplotlib.pyplot as plt
7.   import timeit
8.
9.   # Convert the linguistic variables for the criteria weights
10.  # or the ratings into fuzzy weights and fuzzy decision
11.  # matrix, respectively
12.  def cal(a, b, k):
13.      """ a is the dictionary with the linguistic variables
14.      for the criteria weights (or the linguistic
15.      variables for the ratings), b is the matrix with
16.      the criteria weights (or the ratings), and k is
17.      the number of the decision makers. The output is
18.      the fuzzy decision matrix or the fuzzy weights
19.      of the criteria
20.      """
21.      f = []
22.      for i in range(len(b)):
23.          c = []
24.          for z in range(3):
25.              x = 0
26.              for j in range (k):
27.                  x = x + a[b[i][j]][z]
28.              c.append(round(x / k, 3))
29.          f.append(c)
30.      return asarray(f)
31.
```

```
# Calculate the fuzzy normalized decision matrix
def fndm(a, n, m):
    """ a is the fuzzy decision matrix, n is the number of
    criteria, and m is the number of the alternatives.
    The output is the fuzzy normalized decision matrix
    """
    x = amax(a[:, 2:3])
    f = zeros((n * m, 3))
    for i in range(n * m):
        for j in range(3):
            f[i][j] = round(a[i][j] / x, 3)
    return f

# Calculate the fuzzy weighted normalized decision matrix
def weighted_fndm(a, b, n, m):
    """ a is the fuzzy normalized decision matrix, b is the
    criteria weights, n is the number of criteria, and m
    is the number of the alternatives. The output is
    the fuzzy weighted normalized decision matrix
    """
    f = zeros((n * m, 3))
    z = 0
    for i in range(n * m):
        if i % len(b) == 0:
            z = 0
        else:
            z = z + 1
        for j in range(3):
            f[i][j] = round(a[i][j] * b[z][j], 3)
    return f

# Calculate the distance between two fuzzy triangular
# numbers
def distance(a, b):
    """ a and b are fuzzy triangular numbers. The output is
    their distance
    """
    return sqrt(1/3 * ((a[0] - b[0])**2 + (a[1] - b[1])**2
        + (a[2] - b[2])**2))

# Determine the fuzzy positive ideal solution (FPIS)
def func_dist_fpis(a, n, m):
    """ a is the fuzzy weighted normalized decision matrix,
    n is the number of criteria, and m is the number of
    the alternatives. The output is the ideal
    solution of each criterion
    """
    fpis = ones((3, 1))
    dist_pis = zeros(m)
    p = 0
    for i in range(m):
        for j in range(n):
            dist_pis[i] = dist_pis[i] + distance(a[p + j],
                fpis)
```

```
        p = p + n
    return dist_pis

# Determine the fuzzy negative ideal solution (FNIS)
def func_dist_fnis(a, n, m):
    """ a is the fuzzy weighted normalized decision matrix,
    n is the number of criteria, and m is the number of
    the alternatives. The output is the anti-ideal
    solution of each criterion
    """
    fnis = zeros((3, 1))
    dist_nis = zeros(m)
    p = 0
    for i in range(m):
        for j in range(n):
            dist_nis[i] = dist_nis[i] + distance(a[p + j],
                fnis)
        p = p + n
    return dist_nis

# Fuzzy TOPSIS method: it calls the other functions
def f_topsis(a, b, c, d, n, m, k, pl):
    """ a is the dictionary with the linguistic variables
    for the criteria weights, b is the matrix with the
    importance weights of the criteria, c is a
    dictionary with the linguistic variables for the
    ratings, d is the matrix with all the ratings, n
    is the number of criteria, m is the number of the
    alternatives, and k is the number of the decision
    makers
    """

    # Steps 3 and 4
    fuzzy_weights = cal(a, b, k)
    fuzzy_decision_matrix = cal(c, d, k)
    fuzzy_norm_decision_matrix = fndm(fuzzy_decision_matrix,
        n, m)

    # Step 5
    weighted_fuzzy_norm_decision_matrix = \
        weighted_fndm(fuzzy_norm_decision_matrix,
        fuzzy_weights, n, m)

    # Steps 6 and 7
    a_plus = func_dist_fpis(
       weighted_fuzzy_norm_decision_matrix, n, m)
    a_minus = func_dist_fnis(
       weighted_fuzzy_norm_decision_matrix, n, m)

    # Step 8
    CC = [] # closeness coefficient
    for i in range(m):
        CC.append(round(a_minus[i] / (a_plus[i] +
            a_minus[i]), 3))
```

```
    if pl == 'y':
        q = [i + 1 for i in range(m)]
        plt.plot(q, a_plus, 'p--', color = 'red',
            markeredgewidth = 2, markersize = 8)
        plt.plot(q, a_minus, '*--',  color = 'blue',
            markeredgewidth = 2, markersize = 8)
        plt.plot(q, CC, 'o--', color = 'green',
            markeredgewidth = 2, markersize = 8)
        plt.title('Fuzzy TOPSIS results')
        plt.legend(['Distance from the ideal',
            'Distance from the anti-ideal',
            'Closeness coeficient'])
        plt.xticks(range(m + 2))
        plt.axis([0, m + 1, 0, 3])
        plt.xlabel('Alternatives')
        plt.legend()
        plt.grid(True)
        plt.show()
    return CC

m = 6 # the number of the alternatives
n = 4 # the number of the criteria
k = 3 # the number of the decision makers

# Steps 1 and 2
# Define a dictionary with the linguistic variables for the
# criteria weights
cw = {'VL':[0, 0, 0.1], 'L':[0, 0.1, 0.3],
      'ML':[0.1, 0.3, 0.5], 'M':[0.3, 0.5, 0.7],
      'MH':[0.5, 0.7, 0.9], 'H':[0.7, 0.9, 1],
      'VH':[0.9, 1, 1]}

# Define a dictionary with the linguistic variables for the
# ratings
r = {'VP':[0, 0, 1], 'P':[0, 1, 3], 'MP':[1, 3, 5],
     'F':[3, 5, 7], 'MG':[5, 7, 9], 'G':[7, 9, 10],
     'VG':[9, 10, 10]}

# The matrix with the criteria weights
cdw = [['H', 'VH', 'VH'], ['M', 'H', 'VH'],
        ['M', 'MH', 'ML'], ['H', 'VH', 'MH']]

# The ratings of the six candidate sites by the decision
# makers under all criteria
c1 = [['VG', 'G', 'MG'], ['F', 'MG', 'MG'],
      ['P', 'P', 'MP'], ['G', 'VG', 'G']]
c2 = [['MP', 'F', 'F'], ['F', 'VG', 'G'],
      ['MG', 'VG', 'G'], ['MG', 'F', 'MP']]
c3 = [['MG', 'MP', 'F'], ['MG', 'MG', 'VG'],
      ['MP', 'F', 'F'], ['MP', 'P', 'P']]
c4 = [['MG', 'VG', 'VG'], ['G', 'G', 'VG'],
      ['MG', 'VG', 'G'], ['VP', 'F', 'P']]
c5 = [['VP', 'P', 'G'], ['P', 'VP', 'MP'],
```

```
          ['G', 'G', 'VG'], ['G', 'MG', 'MG']]
c6 = [['F', 'G', 'G'], ['F', 'MP', 'MG'],
          ['VG', 'MG', 'F'], ['P', 'MP', 'F']]

all_ratings = vstack((c1, c2, c3, c4, c5, c6))

# final results
start = timeit.default_timer()
f_topsis(cw, cdw, r, all_ratings, n, m, k, 'n')
stop = timeit.default_timer()
print(stop - start)
print("Closeness coefficient = ",
   f_topsis(cw, cdw, r, all_ratings, n, m, k, 'y'))
```

References

1. Anisseh, M., Piri, F., Shahraki, M. R., & Agamohamadi, F. (2012). Fuzzy extension of TOPSIS model for group decision making under multiple criteria. *Artificial Intelligence Review, 38*(4), 325–338.
2. Bede, B. (2013). *Mathematics of fuzzy sets and fuzzy logic*. Berlin: Springer.
3. Behzadian, M., Otaghsara, S. K., Yazdani, M., & Ignatius, J. (2012). A state-of the-art survey of TOPSIS applications. *Expert Systems with Applications, 39*(17), 13051–13069.
4. Cha, Y., & Jung, M. (2003). Satisfaction assessment of multi-objective schedules using neural fuzzy methodology. *International Journal of Production Research, 41*(8), 1831–1849.
5. Chen, C. T. (2000). Extensions of the TOPSIS for group decision-making under fuzzy environment. *Fuzzy Sets and Systems, 114*, 1–9.
6. Chen, S. H. (1985). Ranking fuzzy numbers with maximizing set and minimizing set. *Fuzzy Sets and Systems, 17*(2), 113–129.
7. Chen, S. J., & Hwang, C. L. (1992). *Fuzzy multiple attribute decision making methods*. Berlin: Springer.
8. Chu, T. C. (2002). Facility location selection using fuzzy TOPSIS under group decisions. *International Journal of Uncertainty, Fuzziness and Knowledge-Based Systems, 10*(6), 687–701.
9. Chu, T. C., & Lin, Y. C. (2003). A fuzzy TOPSIS method for robot selection. *The International Journal of Advanced Manufacturing Technology, 21*(4), 284–290.
10. Diamond, P., & Kloeden, P. (2000). Metric topology of fuzzy numbers and fuzzy analysis. In D. Dubois & H. Prade (Eds.), *Fundamentals of fuzzy sets* (pp. 583–641). New York: Springer US.
11. Dymova, L., Sevastjanov, P., & Tikhonenko, A. (2013). An approach to generalization of fuzzy TOPSIS method. *Information Sciences, 238*, 149–162.
12. Garcia-Cascales, M. S., & Lamata, M. T. (2012). On rank reversal and TOPSIS method. *Mathematical and Computer Modelling, 56*(5), 123–132.
13. Giove, S. (2002). Interval TOPSIS for multicriteria decision making. In *Italian Workshop on Neural Nets* (pp. 56–63). Berlin: Springer.
14. Hwang, C. L., & Yoon, K. (1981). *Multiple attribute decision making: Methods and applications*. Berlin: Springer
15. Izadikhah, M. (2009). Using the Hamming distance to extend TOPSIS in a fuzzy environment. *Journal of Computational and Applied Mathematics, 231*(1), 200–207.
16. Jahan, A., & Edwards, K. L. (2015). A state-of-the-art survey on the influence of normalization techniques in ranking: Improving the materials selection process in engineering design. *Materials & Design, 65*, 335–342.

17. Jahanshahloo, G. R., Lotfi, F. H., & Davoodi, A. R. (2009). Extension of TOPSIS for decision-making problems with interval data: interval efficiency. *Mathematical and Computer Modelling, 49*(5), 1137–1142.
18. Jahanshahloo, G. R., Lotfi, F. H., & Izadikhah, M. (2006). An algorithmic method to extend TOPSIS for decision-making problems with interval data. *Applied Mathematics and Computation, 175*(2), 1375–1384.
19. Jahanshahloo, G. R., Lotfi, F. H., & Izadikhah, M. (2006). Extension of the TOPSIS method for decision-making problems with fuzzy data. *Applied Mathematics and Computation, 181*(2), 1544–1551.
20. Kahraman, C., Kaya, I., Çevik, S., Ates, N. Y., & Gülbay, M. (2008). Fuzzy multi-criteria evaluation of industrial robotic systems using TOPSIS. In C. Kahramn (Ed.), *Fuzzy multi-criteria decision making* (pp. 159–186). New York: Springer US.
21. Kaufmann, A., & Gupta, M. M. (1988). *Fuzzy mathematical models in engineering and management science*. New York: Elsevier Science Inc.
22. Lee, E. S., & Li, R. J. (1988). Comparison of fuzzy numbers based on the probability measure of fuzzy events. *Computers & Mathematics with Applications, 15*(10), 887–896.
23. Lee, K. H. (2006). *First course on fuzzy theory and applications* (Vol. 27). Berlin: Springer Science & Business Media.
24. Li, X., & Chen, X. (2014). Extension of the TOPSIS method based on prospect theory and trapezoidal intuitionistic fuzzy numbers for group decision making. *Journal of Systems Science and Systems Engineering, 23*(2), 231–247.
25. Liang, G. S. (1999). Fuzzy MCDM based on ideal and anti-ideal concepts. *European Journal of Operational Research, 112*(3), 682–691.
26. Liou, T. S., & Wang, M. J. J. (1992). Ranking fuzzy numbers with integral value. *Fuzzy Sets and Systems, 50*(3), 247–255.
27. Mahdavi, I., Mahdavi-Amiri, N., Heidarzade, A., & Nourifar, R. (2008). Designing a model of fuzzy TOPSIS in multiple criteria decision making. *Applied Mathematics and Computation, 206*(2), 607–617.
28. Owen, S. H., & Daskin, M. S. (1998). Strategic facility location: A review. *European Journal of Operational Research, 111*(3), 423–447.
29. Pavlicic, D. (2001). Normalization affects the results of MADM methods. *Yugoslav Journal of Operations Research, 11*(2), 251–265.
30. ReVelle, C. S., & Eiselt, H. A. (2005). Location analysis: a synthesis and survey. *European Journal of Operational Research, 165*(1), 1–19.
31. Shih, H. S., Shyur, H. J., & Lee, E. S. (2007). An extension of TOPSIS for group decision making. *Mathematical and Computer Modelling, 45*(7), 801–813.
32. Tsaur, S. H., Chang, T. Y., & Yen, C. H. (2002). The evaluation of airline service quality by fuzzy MCDM. *Tourism Management, 23*(2), 107–115.
33. Vafaei, N., Ribeiro, R. A., & Camarinha-Matos, L. M. (2015). Importance of data normalization in decision making: Case study with TOPSIS method. In B. Delibasic, F. Dargam, P. Zarate, J.E. Hernendez, S. Liu, R. Ribeiro, I. Linden & J. Papathanasiou (Eds.), *ICDSST 2015 Proceedings - the 1st International Conference on Decision Support Systems Technologies, an EWG-DSS Conference*, Belgrade, Serbia.
34. Wang, T. C., & Lee, H. D. (2009). Developing a fuzzy TOPSIS approach based on subjective weights and objective weights. *Expert Systems with Applications, 36*(5), 8980–8985.
35. Wang, Y. J., & Lee, H. S. (2007). Generalizing TOPSIS for fuzzy multiple-criteria group decision-making. *Computers & Mathematics with Applications, 53*(11), 1762–1772.
36. Wang, Y. M., & Elhag, T. M. (2006). Fuzzy TOPSIS method based on alpha level sets with an application to bridge risk assessment. *Expert Systems with Applications, 31*(2), 309–319.
37. Wang, Y. M., & Luo, Y. (2009). On rank reversal in decision analysis. *Mathematical and Computer Modelling, 49*(5), 1221–1229.
38. Yang, T., & Hung, C. C. (2007). Multiple-attribute decision making methods for plant layout design problem. *Robotics and Computer-Integrated Manufacturing, 23*(1), 126–137.

39. Yue, Z. (2012). Extension of TOPSIS to determine weight of decision maker for group decision making problems with uncertain information. *Expert Systems with Applications, 39*(7), 6343–6350.
40. Zadeh, L. A. (1965). Fuzzy sets. *Information and Control, 8*(3), 338–353.
41. Zadeh, L. A. (1975). The concept of a linguistic variable and its application to approximate reasoning—I. *Information Sciences, 8*(3), 199–249.
42. Zhang, G., & Lu, J. (2003). An integrated group decision-making method dealing with fuzzy preferences for alternatives and individual judgements for selection criteria. *Group Decision and Negotiation, 12*(6), 501–515.
43. Zhao, R., & Govind, R. (1991). Algebraic characteristics of extended fuzzy numbers. *Information Sciences, 54*(1), 103–130.

Chapter 2
VIKOR

2.1 Introduction

According to Opricovic [11], the researcher who originally conceived VIKOR (the acronym is in Serbian: VlseKriterijumska Optimizacija I Kompromisno Resenje, meaning multicriteria optimization and compromise solution), the method has been developed to provide compromise solutions to discrete multiple criteria problems that include non-commensurable and conflicting criteria. It has attracted much attention among researchers and has been applied in various areas (Table 2.1). Its theoretical background is closely related to TOPSIS that was presented in Chapter 1; they are both based on an aggregating function representing the "closeness to the ideal" [15]. VIKOR is considered to be effective in cases where the decision maker cannot be certain how to express his/her preferences coherently and consistently at the initial stages of the system design. Yu [27] and Zeleny [29] provide the setting theory for compromise solutions. Opricovic and Tzeng [15] state that "a compromise solution is a feasible solution, which is closest to the ideal, and a compromise means an agreement established by mutual concessions." As such, the compromise solution can well serve as a basis for negotiations. VIKOR has been successfully applied among other domains to lean tool selection [1], environmental management [2, 3], waste management [10], social sustainability [17], facility location [24], material selection [6], and healthcare management [30]. Like TOPSIS, VIKOR has been combined successfully with many different MCDA methodologies and other techniques. Table 2.1, adopted from [26], presents the distribution of papers on VIKOR by application areas, while Table 2.2, adopted from the same reference, shows techniques that were compared/combined with VIKOR.

Electronic Supplementary Material The online version of this chapter (https://doi.org/10.1007/978-3-319-91648-4_2) contains supplementary material, which is available to authorized users.

J. Papathanasiou, N. Ploskas, *Multiple Criteria Decision Aid*, Springer Optimization and Its Applications 136, https://doi.org/10.1007/978-3-319-91648-4_2

Table 2.1 Distribution of papers on VIKOR by application areas [26]

Area	N	%
Design and manufacturing management	38	19.2
Business and marketing management	35	17.7
Supply chain and logistics management	26	13.1
Environmental resources and energy management	20	10.1
Construction management	9	4.5
Education management	9	4.5
Health care and risk management	7	3.5
Tourism management	7	3.5
Non-application	24	12.1
Other topics	23	11.6
Total	198	100

Table 2.2 Distribution of techniques compared or combined with VIKOR [26]

Techniques compared/combined	N	%	Techniques compared/combined	N	%
Fuzzy approach	89	34.6	Rough-set theory	4	1.6
TOPSIS	30	11.7	BSC	2	0.8
AHP	29	11.3	Genetic algorithm	2	0.8
ANP	27	10.5	Support Vector Machines (SVM)	2	0.8
DEMATEL	17	6.6	WSM	2	0.8
Entropy method	10	3.9	COPRAS	1	0.4
PROMETHEE	9	3.5	QFD	1	0.4
ELECTRE	7	2.7	DEA	1	0.4
Group decision making approach	6	2.3	LINMAP	1	0.4
Delphi method	5	1.9	SERVQUAL	1	0.4
GRA	5	1.9	SWOT	1	0.4
SAW	4	1.6	PCA	1	0.4

2.2 Methodology

The original algorithm, as presented in [15], is composed of five distinct steps and this is the version that we selected to implement here. At a later stage, the method was extended with four additional steps that [12, 16]: (1) provide a stability analysis to determine the weight stability intervals, and (2) include a trade-off analysis. VIKOR has been developed to solve the following problem

$$mco_i \left\{ \left(f_{ij}\left(A_i\right), i = 1, 2, \cdots, m \right), j = 1, 2, \cdots, n \right\} \tag{2.1}$$

where m is the number of feasible alternatives; n is the number of criteria; $A_i = x_1, x_2, \cdots, x_m$ is the ith alternative obtained (generated) with certain values of system variables x; f_{ij} is the value of the jth criterion function for the alternative A_i; and mco denotes the operator of a multicriteria decision making procedure for selecting the best (compromise) alternative in multicriteria sense. The VIKOR algorithm is comprised of five steps as follows:

Step 1. Determine the Best and the Worst Values of All Criteria Functions
Determine the best f_j^* and the worst f_j^- values of all criteria functions

$$f_j^* = \max_i f_{ij}, f_j^- = \min_i f_{ij}, i = 1, 2, \cdots, m, j = 1, 2, \cdots, n \tag{2.2}$$

if the jth function is to be maximized (benefit) and

$$f_j^* = \min_i f_{ij}, f_j^- = \max_i f_{ij}, i = 1, 2, \cdots, m, j = 1, 2, \cdots, n \tag{2.3}$$

if the jth function is to be minimized (cost).

Step 2. Compute the Values S_i and R_i
Compute the values S_i and R_i by the relations

$$S_i = \sum_{j=1}^{n} w_j \left(f_j^* - f_{ij}\right) / \left(f_j^* - f_j^-\right), i = 1, 2, \cdots, m, j = 1, 2, \cdots, n \tag{2.4}$$

$$R_i = \max_j \left[w_j \left(f_j^* - f_{ij}\right) / \left(f_j^* - f_j^-\right)\right], i = 1, 2, \cdots, m, j = 1, 2, \cdots, n \tag{2.5}$$

where w_j is the weight of the jth criterion.

Step 3. Compute the Values Q_i
Compute the values Q_i by the relation

$$Q_i = v\left(S_i - S^*\right) / \left(S^- - S^*\right) + (1 - v)\left(R_i - R^*\right) / \left(R^- - R^*\right), i = 1, 2, \cdots, m \tag{2.6}$$

where $S^* = \min_i S_i$; $S^- = \max_i S_i$; $R^* = \min_i R_i$; $R^- = \max_i R_i$; and v is introduced as a weight for the strategy of the "majority of criteria" (or the "maximum group utility"), whereas $1 - v$ is the weight of the individual regret. These strategies could be compromised by $v = 0.5$, and here v is modified as $v = (n + 1)/2n$ (derived from $v + 0.5(n - 1)/n = 1$) since the criterion (1 of n) related to R is also included in S.

Step 4. Rank the Alternatives
Rank the alternatives, sorting by the values S, R, and Q in ascending order. The results are three ranking lists.

Step 5. Propose a Compromise Solution
Propose as a compromise solution the alternative $[A^{(1)}]$, which is the best ranked by the measure Q (minimum) if the following two conditions are satisfied:

- C1 - Acceptable advantage

$$Q\left(A^{(2)}\right) - Q\left(A^{(1)}\right) \geq DQ \tag{2.7}$$

where $A^{(2)}$ is the second ranked alternative by the measure Q and $DQ = 1/(m-1)$.

- C2 - Acceptable stability in decision making: The alternative $A^{(1)}$ must also be the best ranked by S and/or R. This compromise solution is stable within a decision making process, which could be the strategy of maximum group utility ($v > 0.5$), or "by consensus" ($v \approx 0.5$), or "with veto" ($v < 0.5$). If one of the conditions is not satisfied, then a set of compromise solutions is proposed, which consists of:

  – Alternatives $A^{(1)}$ and $A^{(2)}$ if only the condition C2 is not satisfied, or
  – Alternatives $A^{(1)}, A^{(2)}, \cdots, A^{(l)}$ if the condition C1 is not satisfied; $A^{(l)}$ is determined by the relation $Q(A^{(l)}) - Q(A^{(1)}) < DQ$ for maximum l (the positions of these alternatives are "in closeness").

The original version of the VIKOR method concludes in this step; the next four steps are parts of the extended version.

Step 6. Determine the Weight Stability Interval for Each Criterion
Determine the weight stability interval $\left[w_j^L, w_j^U\right]$ for each criterion, separately, with the initial (given) values of weights. The compromise solution obtained with the initial weights ($w_j, j = 1, 2, \cdots, n$) will be replaced at the highest ranked position if the value of a weight is out of the stability interval. The stability interval is only relevant concerning one-dimensional weighting variations.

Step 7. Determine the Trade-Offs
Determine the trade-offs, $tr_{jk} = \left|\left(D_j w_k\right) / \left(D_k w_j\right)\right|, j = 1, 2, \cdots, n, k = 1, 2, \cdots, n, k \neq j$, where tr_{jk} is the number of units of the jth criterion evaluated the same as one unit of the kth criterion, and $D_j = f_j^* - f_j^-, \forall j$. The index j is given by the decision maker.

Step 8. Adjust the Trade-Offs
The decision maker may give a new value of $tr_{jk}, j = 1, 2, \cdots, n, k = 1, 2, \cdots, n, k \neq j$ if he/she does not agree with the computed values. Then, VIKOR performs a new ranking with the new values of weights $w_k = \left(D_k w_j tr_{jk}\right) / D_j, j = 1, 2, \cdots, n, k = 1, 2, \cdots, n, k \neq j, w_j = 1$ (or the previous values). VIKOR normalizes the weights, with the sum equal to 1. The trade-offs determined in Step 7 could help the decision maker to assess new values, although that task is very difficult.

Step 9. Algorithm Termination
The VIKOR algorithm ends if new values are not given in Step 8.

The results of VIKOR are rankings by S, R, and Q, the proposed compromise solution (one or a set), the weight stability intervals for a single criterion, and the trade-offs introduced by VIKOR.

2.2.1 Numerical Example

Let's assume that we have to solve the same problem as the one presented in Section 1.2.1; that will offer the opportunity to compare the results with those of TOPSIS. Consequently the data input for the experiment are those of Table 1.2. Initially, we calculate the best f_j^* and worst f_j^- values of all criteria functions using Equation (2.2) (Table 2.3). Next, we calculate the difference of each value in the decision matrix from the ideal values of each criterion (Table 2.4). Then, we calculate the values of S_i and R_i using Equations (2.4) and (2.5) (Table 2.5). Finally, we calculate the values Q_i using Equation (2.6) (Table 2.6). The final results are shown in Table 2.6; the first two rows show the values of S_i and R_i, while the third (scenario (a)) shows the values of Q_i for $v = 0.625$ (in this case $v = (4+1)/2*4$, since the criteria are four). Scenarios (b)–(d) show the values of Q_i for different values of v. Figure 2.1 displays the results for $v = 0.625$. The best ranked alternative, with good advantage, is site 5 since it has the minimum measure of $Q(0.000)$ and also according to Step 5 of the algorithm:

- C1 - Acceptable advantage: $Q(A^{(4)}) - Q(A^{(5)}) = 0.271$ and $DQ = 1/(m-1) = 1/5 = 0.2$, therefore, $Q(A^{(4)}) - Q(A^{(5)}) \geq DQ$ (condition satisfied).
- C2 - Acceptable stability in decision making: S_5 (0.080) $\leq S_4$ (0.310) and R_5 (0.080) $\leq R_4$ (0.133); hence, site 5 is better ranked by both S and R (condition satisfied).

The results are the same for the scenarios (b)–(d) and interestingly enough, for the same dataset, TOPSIS produces the same best solution, site 5.

If the initial data in Table 1.2 is slightly modified, i.e., the score of site 5 for the first criterion (investment costs) is reduced from 11 to 1 and the score of site 5 for the second criterion (employment needs) is reduced from 10 to 1, then VIKOR yields the results shown in Table 2.7. In this scenario, site 4 has the minimum measure of $Q(0.000)$, but according to Step 5 of the algorithm:

- C1 - Acceptable advantage: $Q(A^{(6)}) - Q(A^{(4)}) = 0.176$ and $DQ = 1/(m-1) = 1/5 = 0.2$, therefore, $Q(A^{(6)}) - Q(A^{(4)}) \ngeq DQ$ (condition not satisfied).
- C2 - Acceptable stability in decision making: S_4 (0.133) $\leq S_6$ (0.290) and R_4 (0.133) $\leq R_6$ (0.150); hence, site 4 is better ranked by both S and R (condition satisfied).

Table 2.3 Best f_j^* and worst f_j^- values of all criteria functions

	Investment costs	Employment needs	Social impact	Environmental impact
Weight	0.4	0.3	0.1	0.2
Site 1	8	7	2	1
Site 2	5	3	7	5
Site 3	7	5	6	4
Site 4	9	9	7	3
Site 5	11	10	3	7
Site 6	6	9	5	4
f_j^*	11	10	7	7
f_j^-	5	3	2	1
$f_j^* - f_j^-$	6	7	5	6

Table 2.4 Difference from the ideal solution

	Investment costs	Employment needs	Social impact	Environmental impact
Site 1	3	3	5	6
Site 2	6	7	0	2
Site 3	4	5	1	3
Site 4	2	1	0	4
Site 5	0	0	4	0
Site 6	5	1	2	3

Table 2.5 Calculation of S_i and R_i

	Investment costs	Employment needs	Social impact	Environmental impact	S_i	R_i
Site 1	0.200	0.129	0.100	0.200	0.629	0.200
Site 2	0.400	0.300	0.000	0.067	0.767	0.400
Site 3	0.267	0.214	0.020	0.100	0.601	0.267
Site 4	0.133	0.049	0.000	0.133	0.310	0.133
Site 5	0.000	0.000	0.080	0.000	0.080	0.080
Site 6	0.333	0.049	0.040	0.100	0.516	0.333

Condition C1 is not satisfied and then a set of compromise solutions is proposed consisting of both site 4 and site 6 since only condition 1 is not satisfied (site 4 $\approx$ site 6). In this case, $l = 2$, since $Q(A^{(6)}) - Q(A^{(4)}) = 0.176$ ($< DQ$), $Q(A^{(3)}) - Q(A^{(4)}) = 0.253$ ($> DQ$), and the positions of site 4 and site 6 are "in closeness."

Table 2.6 Final results for the facility location problem using initial data

		Site 1	Site 2	Site 3	Site 4	Site 5	Site 6
	S_i	0.629	0.767	0.601	0.310	0.080	0.516
	R_i	0.200	0.400	0.267	0.133	0.080	0.333
(a)	$Q_i(v = 0.625)$	0.640	1.000	0.693	0.271	0.000	0.693
(b)	$Q_i(v = 0.2)$	0.460	1.000	0.618	0.200	0.000	0.760
(c)	$Q_i(v = 0.5)$	0.587	1.000	0.671	0.250	0.000	0.713
(d)	$Q_i(v = 0.8)$	0.714	1.000	0.724	0.301	0.000	0.667

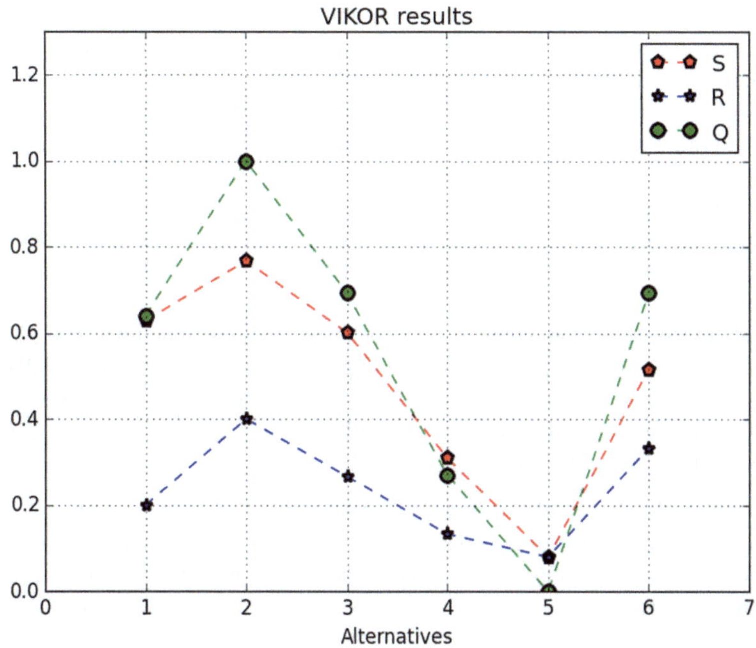

Fig. 2.1 Final results for the facility location problem

Table 2.7 Final results for the facility location problem using modified data

	Site 1	Site 2	Site 3	Site 4	Site 5	Site 6
S_i	0.425	0.492	0.370	0.133	0.780	0.290
R_i	0.200	0.225	0.150	0.133	0.400	0.150
$Q_i(v = 0.625)$	0.376	0.476	0.253	0.000	1.000	0.176

2.2.2 Python Implementation

The file *VIKOR.py* includes a Python implementation of the VIKOR method. Apart from *numpy*, it uses the *matplotlib* library to plot the results and the *timeit* module to time the procedure (lines 5–7). Each step is implemented in its own function and the way to use the function, e.g., the input and output of the function, is embedded as a Python doc string.

More analytically:

- Step 1 (function *best_worst_fij(a, b)*) is in lines 11–23. The inputs are the array with the alternatives performances (variable *a*) and the criteria min/max array (variable *b*). Its output is an array with the best and worst values for all criteria functions (variable *f*).
- Step 2 (function *SR(a, b, c)*) is in lines 26–47. The inputs are the array with the alternatives performances, the array with the best and worst performances, and the criteria min/max array. Its outputs are two vectors with the values of S_i (variable *s*) and R_i (variable *r*).
- Step 3 (function *Q(s, r, n)*) is in lines 50–61. The inputs are the values of S_i and R_i, and the number of criteria. Its output is a vector with the values of Q_i (variable *q*).
- Final results (function *vikor(a, b, c, pl)*) are calculated in lines 64–89. This function calls all the other functions and produces the final result. Variable *a* is the initial decision matrix, *b* is the criteria min/max array, *c* is the criteria weights matrix, and *pl* is whether to plot the results using *matplotlib* or not. Figure 2.1 was produced using the *matplotlib* library.

Using the *timeit* module in lines 102–105 and calling function *vikor* without printing the results (the value of variable *pl* is set to '*n*'), the running time is (an average of 10 runs) 0.0006 s on a Linux machine with an Intel Core i7 at 2.2 GHz CPU and 6 GB RAM. Running the same code with 1,000 sites and four criteria, the execution time is 0.6830 s.

```
1.   # Filename: VIKOR.py
2.   # Description: VIKOR method
3.   # Authors: Papathanasiou, J. & Ploskas, N.
4.
5.   from numpy import *
6.   import matplotlib.pyplot as plt
7.   import timeit
8.
9.   # Step 1: determine the best and worst values for all
10.  # criteria functions
11.  def best_worst_fij(a, b):
12.      """ a is the array with the performances and b is
13.      the criteria min/max array
14.      """
15.      f = zeros((b.shape[0], 2))
```

```
    for i in range(b.shape[0]):
        if b[i] == 'max':
            f[i, 0] = a.max(0)[i]
            f[i, 1] = a.min(0)[i]
        elif b[i] == 'min':
            f[i, 0] = a.min(0)[i]
            f[i, 1] = a.max(0)[i]
    return f

# Step 2: compute the values S_i and R_i
def SR(a, b, c):
    """ a is the array with the performances, b is the
    array with the best and worst performances, and
    c is the criteria min/max array
    """
    s = zeros(a.shape[0])
    r = zeros(a.shape[0])
    for i in range(a.shape[0]):
        k = 0
        o = 0
        for j in range(a.shape[1]):
            k = k + c[j] * (b[j, 0] - a[i, j]) \
                    / (b[j, 0] - b[j, 1])
            u = c[j] * (b[j, 0] - a[i, j]) \
                / (b[j, 0] - b[j, 1])
            if u > o:
                o = u
                r[i] = round(o, 3)
            else:
                r[i] = round(o, 3)
        s[i] = round(k, 3)
    return s, r

# Step 3: compute the values Q_i
def Q(s, r, n):
    """ s is the vector with the S_i values, r is
    the vector with the R_i values, and n is the
    number of criteria
    """
    q = zeros(s.shape[0])
    for i in range(s.shape[0]):
        q[i] = round((((n + 1) / (2 * n)) *
            (s[i] - min(s)) / (max(s) - min(s)) +
            (1 - (n + 1) / (2 * n)) *
            (r[i] - min(r)) / (max(r) - min(r))), 3)
    return q

# VIKOR method: it calls the other functions
def vikor(a, b, c, pl):
    """ a is the decision matrix, b is the criteria
    min/max array, c is the weights matrix, and pl
    is 'y' for plotting the results or any other
    string for not
    """
```

```
    s, r = SR(a, best_worst_fij(a, b), c)
    q = Q(s, r, len(c))
    if pl == 'y':
        e = [i + 1 for i in range(a.shape[0])]
        plt.plot(e, s, 'p--', color = 'red',
           markeredgewidth = 2, markersize = 8)
        plt.plot(e, r, '*--',  color = 'blue',
           markeredgewidth = 2, markersize=8)
        plt.plot(e, q, 'o--', color = 'green',
           markeredgewidth = 2, markersize = 8)
        plt.legend(['S', 'R', 'Q'])
        plt.xticks(range(a.shape[0] + 2))
        plt.axis([0, a.shape[0] + 1, 0,
           max(maximum(maximum(s, r), q)) + 0.3])
        plt.title("VIKOR results")
        plt.xlabel("Alternatives")
        plt.legend()
        plt.grid(True)
        plt.show()
    return s, r, q

# performances of the alternatives
x = array([[8, 7, 2, 1], [5, 3, 7, 5], [7, 5, 6, 4],
    [9, 9, 7, 3], [11, 10, 3, 7], [6, 9, 5, 4]])

# weights of the criteria
w = array([0.4, 0.3, 0.1, 0.2])

# criteria max/min
crit_max_min = array(['max', 'max', 'max', 'max'])

# final results
start = timeit.default_timer()
vikor(x, crit_max_min, w, 'n')
stop = timeit.default_timer()
print(stop - start)
s, r, q = vikor(x, crit_max_min, w, 'y')
print("S = ", s)
print("R = ", r)
print("Q = ", q)
```

2.3 Fuzzy VIKOR for Group Decision Making

2.3.1 Preliminaries: Trapezoidal Fuzzy Numbers

This subsection supplements Section 1.3.1 with additional elements on fuzzy numbers theory to be used for the presentation of fuzzy VIKOR and is based on Lee's work [8]. A trapezoidal fuzzy number A can be defined as $A = (\alpha_1, \alpha_2, \alpha_3, \alpha_4)$ with a membership function determined as follows (Figure 2.2):

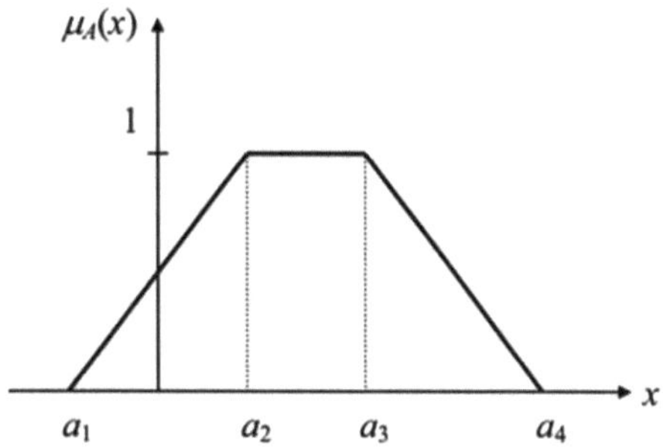

Fig. 2.2 Trapezoidal fuzzy number $A = (\alpha_1, \alpha_2, \alpha_3, \alpha_4)$, adopted from [8]

$$\mu_A(x) = \begin{cases} 0, & x < \alpha_1 \\ \dfrac{x - \alpha_1}{\alpha_2 - \alpha_1}, & \alpha_1 \leq x \leq \alpha_2 \\ 1, & \alpha_2 \leq x \leq \alpha_3 \\ \frac{\alpha_4 - x}{\alpha_4}, & \alpha_3 \leq x \leq \alpha_4 \\ 0, & x > \alpha_4 \end{cases} \tag{2.8}$$

In the case where $\alpha_2 = \alpha_3$, a trapezoidal fuzzy number coincides with a triangular one.

Given a couple of positive trapezoidal fuzzy numbers $A = (\alpha_1, \alpha_2, \alpha_3, \alpha_4)$ and $B = (b_1, b_2, b_3, b_4)$, the result of the addition and subtraction between trapezoidal fuzzy numbers is also a trapezoidal fuzzy number

$$A(+)B = (\alpha_1, \alpha_2, \alpha_3, \alpha_4)(+)(b_1, b_2, b_3, b_4) = (\alpha_1 + b_1, \alpha_2 + b_2, \alpha_3 + b_3, \alpha_4 + b_4) \tag{2.9}$$

and

$$A(-)B = (\alpha_1, \alpha_2, \alpha_3, \alpha_4)(-)(b_1, b_2, b_3, b_4) = (\alpha_1 - b_1, \alpha_2 - b_2, \alpha_3 - b_3, \alpha_4 - b_4) \tag{2.10}$$

As for multiplication, division, and inverse, the result is not a trapezoidal fuzzy number.

The Center of Area (COA) method, which is used for calculating the crisp value of a fuzzy number in the next subsection, can be expressed using the relation [21]

$$\begin{aligned} \text{defuzz}\left(\widetilde{A}\right) &= \frac{\int x \cdot \mu(x)dx}{\int \mu(x)dx} = \frac{\int_{\alpha_1}^{\alpha_2} \left(\frac{x-\alpha_1}{\alpha_2-\alpha_1}\right) \cdot xdx + \int_{\alpha_2}^{\alpha_3} xdx + \int_{\alpha_3}^{\alpha_4} \left(\frac{\alpha_4-x}{\alpha_4-\alpha_3}\right) \cdot xdx}{\int_{\alpha_1}^{\alpha_2} \left(\frac{x-\alpha_1}{\alpha_2-\alpha_1}\right) dx + \int_{\alpha_2}^{\alpha_3} dx + \int_{\alpha_3}^{\alpha_4} \left(\frac{\alpha_4-x}{\alpha_4-\alpha_3}\right) dx} \\ &= \frac{-\alpha_1\alpha_2 + \alpha_3\alpha_4 + \frac{1}{3}\left(\alpha_4 - \alpha_3\right)^2 - \frac{1}{3}\left(\alpha_2 - \alpha_1\right)^2}{-\alpha_1 - \alpha_2 + \alpha_3 + \alpha_4} \end{aligned} \tag{2.11}$$

2.3.2 Fuzzy VIKOR Methodology

This section presents a fuzzy extension of VIKOR that is based on the methodology proposed by Sanayei et al. [21]; they use trapezoidal fuzzy numbers and in their paper, they focus on the supplier selection problem but the methodology can easily be applied in a broader scope as well. The fuzzy version of TOPSIS presented in Section 1.3.2 was based on triangular fuzzy numbers; studies of fuzzy VIKOR with this kind of fuzzy numbers have been implemented in the past by Opricovic [13], Rostamzadeh et al. [20], Chen and Wang [4], and Wan et al. [25]. Shemshadi et al. [23], Ju and Wang [7], and Yucenur and Demirel [28] have worked with trapezoidal fuzzy numbers. Other studies in fuzzy VIKOR include Opricovic and Tzeng [14], Park et al. [18], Liu et al. [9], and Devi [5], while Sayadi et al. [22] extended VIKOR with interval numbers.

The steps of the procedure proposed by Sanayei et al. [21] are:

Step 1. Identify the Objectives of the Decision Making Process and Define the Problem Scope
In this step, the decision goals and the scope of the problem are defined. Then, the objectives of the decision making process are identified.

Step 2. Arrange the Decision Making Group and Define and Describe a Finite Set of Relevant Attributes
We form a group of decision makers to identify the criteria and their evaluation scales.

Step 3. Identify the Appropriate Linguistic Variables
Choose the appropriate linguistic variables for the importance weight of the criteria and the linguistic ratings for alternatives with respect to the criteria.

Step 4. Pull the Decision Makers' Opinions to Get the Aggregated Fuzzy Weight of Criteria and Aggregated Fuzzy Rating of Alternatives, and Construct a Fuzzy Decision Matrix
Let the fuzzy rating and importance weight of the kth decision maker be $\widetilde{x}_{ijk} = \left(\widetilde{x}_{ijk1}, \widetilde{x}_{ijk2}, \widetilde{x}_{ijk3}, \widetilde{x}_{ijk4}\right)$ and $\widetilde{w}_{ijk} = \left(\widetilde{w}_{ijk1}, \widetilde{w}_{ijk2}, \widetilde{w}_{ijk3}, \widetilde{w}_{ijk4}\right)$, respectively, where $i = 1, 2, \cdots, m$ and $j = 1, 2, \cdots, n$. Hence, the aggregated fuzzy ratings $(\widetilde{x}_{ij})$ of alternatives with respect to each criterion can be calculated as

$$\widetilde{x}_{ij} = \left(x_{ij1}, x_{ij2}, x_{ij3}, x_{ij4}\right) \tag{2.12}$$

where

$$\begin{aligned} x_{ij1} &= \min_k \left\{x_{ijk1}\right\}, x_{ij2} = \frac{1}{K}\sum_{k=1}^{K} x_{ijk2}, \\ x_{ij3} &= \frac{1}{K}\sum_{k=1}^{K} x_{ijk3}, x_{ij4} = \max_k \left\{x_{ijk4}\right\} \end{aligned} \tag{2.13}$$

The aggregated fuzzy weights ($\widetilde{w}_j$) of each criterion can be calculated as

$$\widetilde{w}_j = \left(w_{j1}, w_{j2}, w_{j3}, w_{j4}\right) \tag{2.14}$$

where

$$w_{j1} = \min_k \left\{w_{jk1}\right\}, w_{j2} = \frac{1}{K}\sum_{k=1}^{K} w_{jk2}, \\ w_{j3} = \frac{1}{K}\sum_{k=1}^{K} w_{jk3}, w_{ij4} = \max_k \left\{w_{jk4}\right\} \tag{2.15}$$

The problem can be concisely expressed in matrix format as follows:

$$\widetilde{D} = \begin{bmatrix} \widetilde{x}_{11} & \widetilde{x}_{12} & \cdots & \widetilde{x}_{1n} \\ \widetilde{x}_{21} & \widetilde{x}_{22} & \cdots & \widetilde{x}_{2n} \\ \vdots & \vdots & \ddots & \vdots \\ \widetilde{x}_{m1} & \widetilde{x}_{m2} & \cdots & \widetilde{x}_{mn} \end{bmatrix} \tag{2.16}$$

and the vector of the criteria weights as

$$\widetilde{W} = [\widetilde{w}_1, \widetilde{w}_2, \cdots, \widetilde{w}_n] \tag{2.17}$$

where $\widetilde{x}_{ij}$ and $\widetilde{w}_j, i = 1, 2, \cdots, m, j = 1, 2, \cdots, n$, are linguistic variables according to Step 3. They can be approximated by the trapezoidal fuzzy numbers $\widetilde{x}_{ij} = \left(x_{ij1}, x_{ij2}, x_{ij3}, x_{ij4}\right)$ and $\widetilde{w}_j = \left(w_{j1}, w_{j2}, w_{j3}, w_{j4}\right)$.

Step 5. Defuzzify the Fuzzy Decision Matrix and Fuzzy Weight of Each Criterion into Crisp Values
Deffuzzify the fuzzy decision matrix and fuzzy weight of each criterion into crisp values using COA defuzzification relation (Equation (2.11)). Other methods have also been proposed for this step (for a review, see [19]).

Step 6. Determine the Best and the Worst Values of All Criteria Functions
Determine the best f_j^* and the worst f_j^- values of all criteria functions

$$f_j^* = \max_i f_{ij}, f_j^- = \min_i f_{ij}, i = 1, 2, \cdots, m, j = 1, 2, \cdots, n \tag{2.18}$$

if the jth function is to be maximized (benefit) and

$$f_j^* = \min_i f_{ij}, f_j^- = \max_i f_{ij}, i = 1, 2, \cdots, m, j = 1, 2, \cdots, n \tag{2.19}$$

if the jth function is to be minimized (cost).

Step 7. Compute the Values S_i and R_i
Compute the values S_i and R_i by the relations

$$S_i = \sum_{j=1}^{n} w_j \left(f_j^* - f_{ij}\right) / \left(f_j^* - f_j^-\right), i = 1, 2, \cdots, m, j = 1, 2, \cdots, n \tag{2.20}$$

$$R_i = \max_j \left[w_j \left(f_j^* - f_{ij}\right) / \left(f_j^* - f_j^-\right)\right], i = 1, 2, \cdots, m, j = 1, 2, \cdots, n \tag{2.21}$$

Step 8. Compute the Values Q_i
Compute the values Q_i by the relation

$$Q_i = v\left(S_i - S^*\right) / \left(S^- - S^*\right) + (1 - v)\left(R_i - R^*\right) / \left(R^- - R^*\right), i = 1, 2, \cdots, m \tag{2.22}$$

where $S^* = \min_i S_i$; $S^- = \max_i S_i$; $R^* = \min_i R_i$; $R^- = \max_i R_i$; and v is introduced as a weight for the strategy of the "maximum group utility," whereas $1 - v$ is the weight of the individual regret.

Step 9. Rank the Alternatives
Rank the alternatives, sorting by the values S, R, and Q in ascending order. The results are three ranking lists.

Step 10. Propose a Compromise Solution
Propose as a compromise solution the alternative $[A^{(1)}]$, which is the best ranked by the measure Q (minimum) if the following two conditions are satisfied:

- C1 - Acceptable advantage

$$Q\left(A^{(2)}\right) - Q\left(A^{(1)}\right) \geq DQ \tag{2.23}$$

  where $A^{(2)}$ is the second ranked alternative by the measure Q and $DQ = 1/(m - 1)$.
- C2 - Acceptable stability in decision making: The alternative $A^{(1)}$ must also be the best ranked by S and/or R. This compromise solution is stable within a decision making process, which could be the strategy of maximum group utility ($v > 0.5$), or "by consensus" ($v \approx 0.5$), or "with veto" ($v < 0.5$). If one of the conditions is not satisfied, then a set of compromise solutions is proposed, which consists of:
  - Alternatives $A^{(1)}$ and $A^{(2)}$ if only the condition C2 is not satisfied, or
  - Alternatives $A^{(1)}, A^{(2)}, \cdots, A^{(l)}$ if the condition C1 is not satisfied; $A^{(l)}$ is determined by the relation $Q(A^{(l)}) - Q(A^{(1)}) < DQ$ for maximum l (the positions of these alternatives are "in closeness").

2.3.3 Numerical Example

The problem in hand is the same as in the previous examples, the facility location problem. In this case, the importance weights of the qualitative criteria and the ratings are considered as linguistic variables expressed in positive trapezoidal fuzzy numbers, as shown in Tables 2.8 and 2.9; they are also considered to be evaluated by decision makers that are experts on the field. These evaluations are in Tables 2.10 and 2.11.

Initially, we aggregate the weights of criteria to get the aggregated fuzzy weights and the ratings to calculate the fuzzy decision matrix using Equations (2.12)–(2.17) (Table 2.12). Then, we deffuzzify the fuzzy decision matrix and fuzzy weight of each criterion into crisp values using Equation (2.11) (Table 2.13). Next, we calculate the best f_j^* and worst f_j^- values of all criteria functions using Equations (2.18) and (2.19) (Table 2.14). Then, we calculate the values of S_i and R_i using Equations (2.20) and (2.21) (Table 2.15), respectively. Finally, we calculate the values Q_i using Equation (2.22) (Table 2.16). The final results are shown in Table 2.16 and Figure 2.3. The best ranked alternative by the measure Q is site 1, which also satisfies the conditions C1 and C2.

Table 2.8 Linguistic variables for the criteria

Linguistic variables for the importance weight of each criterion	
Very low (VL)	(0, 0, 0.1, 0.2)
Low (L)	(0.1, 0.2, 0.2, 0.3)
Medium low (ML)	(0.2, 0.3, 0.4, 0.5)
Medium (M)	(0.4, 0.5, 0.5, 0.6)
Medium high (MH)	(0.5, 0.6, 0.7, 0.8)
High (H)	(0.7, 0.8, 0.8, 0.9)
Very high (VH)	(0.8, 0.9, 1.0, 1.0)

Table 2.9 Linguistic variables for the ratings

Linguistic variables for the ratings	
Very poor (VP)	(0.0, 0.0, 0.1, 0.2)
Poor (P)	(0.1, 0.2, 0.2, 0.3)
Medium poor (MP)	(0.2, 0.3, 0.4, 0.5)
Fair (F)	(0.4, 0.5, 0.5, 0.6)
Medium good (MG)	(0.5, 0.6, 0.7, 0.8)
Good (G)	(0.7, 0.8, 0.8, 0.9)
Very good (VG)	(0.8, 0.9, 1.0, 1.0)

Table 2.10 The importance weight of the criteria for each decision maker

	D_1	D_2	D_3
Investment costs	H	VH	VH
Employment needs	M	H	VH
Social impact	M	MH	ML
Environmental impact	H	VH	MH

Table 2.11 The ratings of the six sites by the three decision makers for the four criteria

		Decision makers		
Criteria	Candidate sites	D_1	D_2	D_3
Investment costs	Site 1	VG	G	MG
	Site 2	MP	F	F
	Site 3	MG	MP	F
	Site 4	MG	VG	VG
	Site 5	VP	P	G
	Site 6	F	G	G
Employment needs	Site 1	F	MG	MG
	Site 2	F	VG	G
	Site 3	MG	MG	VG
	Site 4	G	G	VG
	Site 5	P	VP	MP
	Site 6	F	MP	MG
Social impact	Site 1	P	P	MP
	Site 2	MG	VG	G
	Site 3	MP	F	F
	Site 4	MG	VG	G
	Site 5	G	G	VG
	Site 6	VG	MG	F
Environmental impact	Site 1	G	VG	G
	Site 2	MG	F	MP
	Site 3	MP	P	P
	Site 4	VP	F	P
	Site 5	G	MG	MG
	Site 6	P	MP	F

2.3.4 *Python Implementation*

The file *FuzzyVIKOR.py* includes a Python implementation of the Fuzzy VIKOR method. Similar to the implementation of VIKOR, it uses the *matplotlib* library to plot the results and the *timeit* module to time the procedure (lines 6–7). Each step is implemented in its own function and the way to use the function, e.g., the input of the function, is embedded as a Python doc string.

More analytically:

- Steps 1–3 of the fuzzy VIKOR procedure are in lines 159–193. Dictionaries *cw* (lines 159–162) and *r* (lines 166–169) correspond to Tables 2.8 and 2.9, respectively; they are the definitions of the linguistic variables used for the criteria weights and the ratings. Python list *cdw* (lines 172–173) is about the importance weight of the criteria (Table 2.10) and the ratings of the six candidate sites by the three decision makers are each in its own list, named *c1*, *c2*, $\cdots$, *c6*

Table 2.12 Fuzzy decision matrix and fuzzy weights

	Investment costs	Employment needs	Social impact	Environmental impact
Site 1	(0.500, 0.767, 0.833, 1.000)	(0.400, 0.567, 0.633, 0.800)	(0.100, 0.233, 0.267, 0.500)	(0.700, 0.833, 0.867, 1.000)
Site 2	(0.200, 0.433, 0.467, 0.600)	(0.400, 0.733, 0.767, 1.000)	(0.500, 0.767, 0.833, 1.000)	(0.200, 0.467, 0.533, 0.800)
Site 3	(0.200, 0.467, 0.533, 0.800)	(0.500, 0.700, 0.800, 1.000)	(0.200, 0.433, 0.467, 0.600)	(0.100, 0.233, 0.267, 0.500)
Site 4	(0.500, 0.800, 0.900, 1.000)	(0.700, 0.833, 0.867, 1.000)	(0.500, 0.767, 0.833, 1.000)	(0.000, 0.233, 0.267, 0.600)
Site 5	(0.000, 0.333, 0.367, 0.900)	(0.000, 0.167, 0.233, 0.500)	(0.700, 0.833, 0.867, 1.000)	(0.500, 0.667, 0.733, 0.900)
Site 6	(0.400, 0.700, 0.700, 0.900)	(0.200, 0.467, 0.533, 0.800)	(0.400, 0.667, 0.733, 1.000)	(0.100, 0.333, 0.367, 0.600)
Weight	(0.700, 0.867, 0.933, 1.000)	(0.400, 0.733, 0.767, 1.000)	(0.200, 0.467, 0.533, 0.800)	(0.500, 0.767, 0.833, 1.000)

Table 2.13 Crisp values for decision matrix and weight of each criterion

	Investment costs	Employment needs	Social impact	Environmental impact
Site 1	0.769	0.769	0.789	0.850
Site 2	0.600	0.500	0.850	0.700
Site 3	0.282	0.500	0.769	0.667
Site 4	0.850	0.750	0.282	0.500
Site 5	0.418	0.418	0.415	0.700
Site 6	0.718	0.282	0.231	0.350
Weight	0.870	0.718	0.500	0.769

Table 2.14 Best f_j^* and worst f_j^- values of all criteria functions

	Investment costs	Employment needs	Social impact	Environmental impact
Weight	0.870	0.718	0.500	0.769
Site 1	0.769	0.769	0.789	0.850
Site 2	0.600	0.500	0.850	0.700
Site 3	0.282	0.500	0.769	0.667
Site 4	0.850	0.750	0.282	0.500
Site 5	0.418	0.418	0.415	0.700
Site 6	0.718	0.282	0.231	0.350
f_j^*	0.850	0.769	0.850	0.850
f_j^-	0.282	0.282	0.231	0.350
$f_j^* - f_j^-$	0.568	0.487	0.619	0.500

Table 2.15 Calculation of S_i and R_i

	S_i	R_i
Site 1	0.173	0.124
Site 2	1.010	0.397
Site 3	1.613	0.87
Site 4	1.025	0.538
Site 5	1.761	0.662
Site 6	2.189	0.769

Table 2.16 Final results for the facility location problem

	Site 1	Site 2	Site 3	Site 4	Site 5	Site 6
S_i	0.173	1.010	1.613	1.025	1.761	2.189
R_i	0.124	0.397	0.870	0.538	0.662	0.769
$Q_i(v = 0.625)$	0.000	0.397	0.821	0.472	0.763	0.949

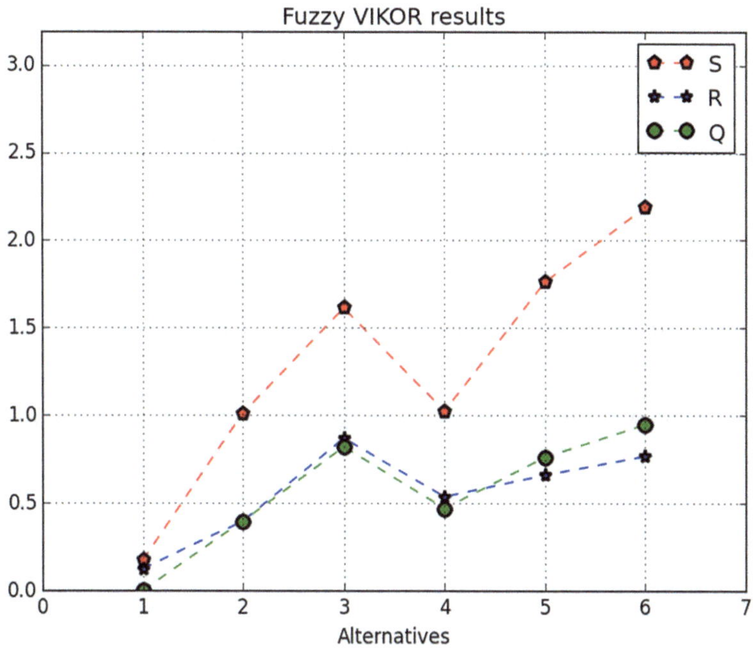

Fig. 2.3 Final results for the facility location problem

(lines 177–188, Table 2.11). These lists are concatenated into one list in line 190. The criteria min/max array is defined in line 193.

- Step 4 is in lines 12–38. Function *agg_fuzzy_value* takes as input a dictionary with the linguistic variables for the importance weight of each criterion or the linguistic variables for the ratings (a), the matrix with the importance weights of the criteria or the ratings by the decision makers (b), and the number of the decision makers (k). Array f is the output of this function and stores the fuzzy decision matrix or fuzzy weights (Table 2.12).
- Step 5 is in lines 42–49. Function *defuzz* takes as input a fuzzy trapezoidal number and returns a crisp value using Equation (2.11) (Table 2.13).
- Step 6 (function *best_worst_fij(a, b)*) is in lines 53–65. The inputs are the array with the alternatives performances (variable a) and the criteria min/max array (variable b). Its output is an array with the best and worst values for all criteria functions (variable f, Table 2.14).
- Step 7 (function *SR(a, b, c)*) is in lines 68–89. The inputs are the array with the alternatives performances, the array with the best and worst performances, and the criteria min/max array. Its outputs are two vectors with the values of S_i (variable s) and R_i (variable r, Table 2.15)
- Step 8 (function *Q(s, r, n)*) is in lines 92–103. The inputs are the values of S_i and R_i, and the number of criteria. Its output is a vector with the values of Q_i (variable q, Table 2.16).

- Final results (function *f_vikor(a, b, c, d, e, n, m, k, pl)*) are calculated in lines 105–150. This function calls all the other functions and produces the final result. Variable *a* is the dictionary with the linguistic variables for the criteria weights, variable *b* is the matrix with the importance weights of the criteria, variable *c* is a dictionary with the linguistic variables for the ratings, variable *d* is the matrix with all the ratings, variable *e* is the criteria max_min array, variable *n* is the number of criteria, variable *m* is the number of the alternatives, variable *k* is the number of the decision makers, and variable *pl* is whether to plot the results using *matplotlib* or not. Figure 2.3 was produced using the *matplotlib* library.

Using the *timeit* module in lines 196–200 and calling function *f_vikor* without printing the results (the value of variable *pl* is set to '*n*'), the running time is (an average of 10 runs) 0.0016 s on a Linux machine with an Intel Core i7 at 2.2 GHz CPU and 6 GB RAM. Running the same code with 1,000 sites, four criteria and three decision makers, the execution time is 0.7164 s.

```
1.    # Filename: FuzzyVIKOR.py
2.    # Description: Fuzzy VIKOR method
3.    # Authors: Papathanasiou, J. & Ploskas, N.
4.
5.    from numpy import *
6.    import matplotlib.pyplot as plt
7.    import timeit
8.
9.    # Step 4: Convert the linguistic variables for the criteria
10.   # weights or the ratings into fuzzy weights and fuzzy
11.   # decision matrix, respectively
12.   def agg_fuzzy_value(a, b, k):
13.       """ a is the dictionary with the linguistic variables
14.       for the criteria weights (or the linguistic
15.       variables for the ratings), b is the matrix with
16.       the criteria weights (or the ratings), and k is
17.       the number of the decision makers. The output
18.       is the fuzzy decision matrix or the fuzzy
19.       weights of the criteria
20.       """
21.       f = zeros((len(b), 4))
22.       for j in range(len(b)):
23.           k0 = a[b[j][0]][0]
24.           k1 = 0
25.           k2 = 0
26.           k3 = a[b[j][0]][3]
27.           for i in range (len(b[1])):
28.               if k0 > a[b[j][i]][0]:
29.                   k0 = a[b[j][i]][0]
30.               k1 = k1 + a[b[j][i]][1]
31.               k2 = k2 + a[b[j][i]][2]
32.               if k3 < a[b[j][i]][3]:
33.                   k3 = a[b[j][i]][3]
34.           f[j][0] = round(k0, 3)
```

```
            f[j][1] = round(k1 / k, 3)
            f[j][2] = round(k2 / k, 3)
            f[j][3] = round(k3, 3)
        return f

    # Step 5: Deffuzify a trapezoidal fuzzy number into
    # a crisp value
    def defuzz(a):
        """ a is a trapezoidal matrix. The output is a
        crisp value
        """
        return (-a[0] * a[1] + a[2] * a[3] +
            1 / 3 * (a[3] - a[2])**2 -
            1 / 3 * (a[1] - a[0])**2) \
            / (-a[0] - a[1] + a[2] + a[3])

    # Step 6: Determine the best and worst values for all
    # criteria functions
    def best_worst_fij(a, b):
        """ a is the array with the performances and b is
        the criteria min/max array
        """
        f = zeros((b.shape[0], 2))
        for i in range(b.shape[0]):
            if b[i] == 'max':
                f[i, 0] = a.max(0)[i]
                f[i, 1] = a.min(0)[i]
            elif b[i] == 'min':
                f[i, 0] = a.min(0)[i]
                f[i, 1] = a.max(0)[i]
        return f

    # Step 7: Compute the values S_i and R_i
    def SR(a, b, c):
        """ a is the array with the performances, b is the
        array with the best and worst performances, and c
        is the criteria min/max array
        """
        s = zeros(a.shape[0])
        r = zeros(a.shape[0])
        for i in range(a.shape[0]):
            k = 0
            o = 0
            for j in range(a.shape[1]):
                k = k + c[j] * (b[j, 0] - a[i, j]) \
                        / (b[j, 0] - b[j, 1])
                u = c[j] * (b[j, 0] - a[i, j]) \
                    / (b[j, 0] - b[j, 1])
                if u > o:
                    o = u
                    r[i] = round(o, 3)
                else:
                    r[i] = round(o, 3)
            s[i] = round(k, 3)
```

```
    return s, r

# Step 8: compute the values Q_i
def Q(s, r, n):
    """ s is the vector with the S_i values, r is
    the vector with the R_i values, and n is the
    number of criteria
    """
    q = zeros(s.shape[0])
    for i in range(s.shape[0]):
        q[i] = round((((n + 1) / (2 * n)) *
            (s[i] - min(s)) / (max(s) - min(s)) +
            (1 - (n + 1) / (2 * n)) *
            (r[i] - min(r)) / (max(r) - min(r))), 3)
    return q

def f_vikor(a, b, c, d, e, n, m, k, pl):
    """ a is the dictionary with the linguistic variables
    for the criteria weights, b is the matrix with the
    importance weights of the criteria, c is a
    dictionary with the linguistic variables for the
    ratings, d is the matrix with all the ratings, e
    is the criteria max_min array, n is the number
    of criteria, m is the number of the alternatives,
    k is the number of the decision makers, and pl
    is 'y' for plotting the results
    """

    w = agg_fuzzy_value(a, b, k)
    f_rdm_all = agg_fuzzy_value(c, d, k)
    crisp_weights = zeros(n)
    for i in range(n):
        crisp_weights[i] = round(defuzz(w[i]), 3)
    crisp_alternative_ratings = zeros((m, n))
    k = 0
    for i in range(n):
        for j in range(m):
            crisp_alternative_ratings[j][i] = \
                round(defuzz(f_rdm_all[k]), 3)
            k = k + 1
    s, r = SR(crisp_alternative_ratings,
        best_worst_fij(crisp_alternative_ratings, e),
        crisp_weights)
    q = Q(s, r, len(w))
    if pl == 'y':
        h = [i + 1 for i in range(m)]
        plt.plot(h, s, 'p--', color = 'red',
           markeredgewidth = 2, markersize=8)
        plt.plot(h, r, '*--',  color = 'blue',
           markeredgewidth = 2, markersize = 8)
        plt.plot(h, q, 'o--', color = 'green',
           markeredgewidth = 2, markersize = 8)
        plt.legend(['S', 'R', 'Q'])
        plt.xticks(range(m + 2))
```

```
        plt.axis([0, m + 1, 0,
           max(maximum(maximum(s, r), q)) + 1])
        plt.title("Fuzzy VIKOR results")
        plt.xlabel("Alternatives")
        plt.legend()
        plt.grid(True)
        plt.show()
    return s, r, q

m = 6 # the number of the alternatives
n = 4 # the number of the criteria
k = 3 # the number of the decision makers

# Steps 1, 2 and 3
# Define a dictionary with the linguistic variables for the
# criteria weights
cw = {'VL':[0, 0, 0.1, 0.2], 'L':[0.1, 0.2, 0.2, 0.3],
    'ML':[0.2, 0.3, 0.4, 0.5], 'M':[0.4, 0.5, 0.5, 0.6],
    'MH':[0.5, 0.6, 0.7, 0.8], 'H':[0.7, 0.8, 0.8, 0.9],
    'VH':[0.8, 0.9, 1, 1]}

# Define a dictionary with the linguistic variables for the
# ratings
r = {'VP':[0.0, 0.0, 0.1, 0.2], 'P':[0.1, 0.2, 0.2, 0.3],
    'MP':[ 0.2, 0.3, 0.4, 0.5], 'F':[0.4, 0.5, 0.5, 0.6],
    'MG':[0.5, 0.6, 0.7, 0.8], 'G':[0.7, 0.8, 0.8, 0.9],
    'VG':[0.8, 0.9, 1.0, 1.0]}

# The matrix with the criteria weights
cdw = [['H', 'VH', 'VH'], ['M', 'H', 'VH'],
   ['M', 'MH', 'ML'], ['H', 'VH', 'MH']]

# The ratings of the six candidate sites by the decision
# makers under all criteria
c1 = [['VG', 'G', 'MG'], ['F', 'MG', 'MG'],
   ['P', 'P', 'MP'], ['G', 'VG', 'G']]
c2 = [['MP', 'F', 'F'], ['F', 'VG', 'G'],
   ['MG', 'VG', 'G'], ['MG', 'F', 'MP']]
c3 = [['MG', 'MP', 'F'], ['MG', 'MG', 'VG'],
   ['MP', 'F', 'F'], ['MP', 'P', 'P']]
c4 = [['MG', 'VG', 'VG'], ['G', 'G', 'VG'],
   ['MG', 'VG', 'G'], ['VP', 'F', 'P']]
c5 = [['VP', 'P', 'G'], ['P', 'VP', 'MP'],
   ['G', 'G', 'VG'], ['G', 'MG', 'MG']]
c6 = [['F', 'G', 'G'], ['F', 'MP', 'MG'],
   ['VG', 'MG', 'F'], ['P', 'MP', 'F']]

all_ratings = vstack((c1, c2, c3, c4, c5, c6))

# criteria max/min array
crit_max_min = array(['max', 'max', 'max', 'max'])

# final results
start = timeit.default_timer()
```

```
f_vikor(cw, cdw, r, all_ratings, crit_max_min, n, m,
    k, 'n')
stop = timeit.default_timer()
print(stop - start)
s, r, q = f_vikor(cw, cdw, r, all_ratings,
    crit_max_min, n, m, k, 'y')
print("S = ", s)
print("R = ", r)
print("Q = ", q)
```

References

1. Anvari, A., Zulkifli, N., & Arghish, O. (2014). Application of a modified VIKOR method for decision-making problems in lean tool selection. *The International Journal of Advanced Manufacturing Technology, 71*(5–8), 829–841.
2. Chang, C. L., & Hsu, C. H. (2009). Multi-criteria analysis via the VIKOR method for prioritizing land-use restraint strategies in the Tseng-Wen reservoir watershed. *Journal of Environmental Management, 90*(11), 3226–3230.
3. Chang, C. L., & Hsu, C. H. (2011). Applying a modified VIKOR method to classify land subdivisions according to watershed vulnerability. *Water Resources Management, 25*(1), 301–309.
4. Chen, L. Y., & Wang, T. C. (2009). Optimizing partners' choice in IS/IT outsourcing projects: The strategic decision of fuzzy VIKOR. *International Journal of Production Economics, 120*(1), 233–242.
5. Devi, K. (2011). Extension of VIKOR method in intuitionistic fuzzy environment for robot selection. *Expert Systems with Applications, 38*(11), 14163–14168.
6. Jahan, A., Mustapha, F., Ismail, M. Y., Sapuan, S. M., & Bahraminasab, M. (2011). A comprehensive VIKOR method for material selection. *Materials & Design, 32*(3), 1215–1221.
7. Ju, Y., & Wang, A. (2013). Extension of VIKOR method for multi-criteria group decision making problem with linguistic information. *Applied Mathematical Modelling, 37*(5), 3112–3125.
8. Lee, K. H. (2006). *First course on fuzzy theory and applications* (Vol. 27). Berlin: Springer Science & Business Media.
9. Liu, H. C., Liu, L., Liu, N., & Mao, L. X. (2012). Risk evaluation in failure mode and effects analysis with extended VIKOR method under fuzzy environment. *Expert Systems with Applications, 39*(17), 12926–12934.
10. Liu, H. C., You, J. X., Fan, X. J., & Chen, Y. Z. (2014). Site selection in waste management by the VIKOR method using linguistic assessment. *Applied Soft Computing, 21*, 453–461.
11. Opricovic, S. (1998). *Multicriteria optimization of civil engineering systems*. PhD Thesis, Faculty of Civil Engineering, Belgrade.
12. Opricovic, S. (2009). A compromise solution in water resources planning. *Water Resources Management, 23*(8), 1549–1561.
13. Opricovic, S. (2011). Fuzzy VIKOR with an application to water resources planning. *Expert Systems with Applications, 38*(10), 12983–12990.
14. Opricovic, S., & Tzeng, G. H. (2002). Multicriteria planning of post-earthquake sustainable reconstruction. *Computer-Aided Civil and Infrastructure Engineering, 17*(3), 211–220.
15. Opricovic, S., & Tzeng, G. H. (2004). Compromise solution by MCDM methods: A comparative analysis of VIKOR and TOPSIS. *European Journal of Operational Research, 156*(2), 445–455.
16. Opricovic, S., & Tzeng, G. H. (2007). Extended VIKOR method in comparison with outranking methods. *European Journal of Operational Research, 178*(2), 514–529.

17. Papathanasiou, J., Ploskas, N., Bournaris, T., & Manos, B. (2016). A decision support system for multiple criteria alternative ranking using TOPSIS and VIKOR: A case study on social sustainability in agriculture. In S. Liu et al. (Eds.), *Decision support systems vi – decision support systems addressing sustainability & societal challenges. Lecture notes in business information processing* (Vol. 250, pp. 3–15). New York: Springer.
18. Park, J. H., Cho, H. J., & Kwun, Y. C. (2011). Extension of the VIKOR method for group decision making with interval-valued intuitionistic fuzzy information. *Fuzzy Optimization and Decision Making, 10*(3), 233–253.
19. Ploskas, N., Papathanasiou, J., & Tsaples, G. (2017). Implementation of an extended fuzzy VIKOR method based on triangular and trapezoidal fuzzy linguistic variables and alternative deffuzification techniques. In I. Linden et al. (Eds.), *Decision support systems vii – data, information and knowledge in decision support systems. Lecture notes in business information processing*. New York: Springer.
20. Rostamzadeh, R., Govindan, K., Esmaeili, A., & Sabaghi, M. (2015). Application of fuzzy VIKOR for evaluation of green supply chain management practices. *Ecological Indicators, 49*, 188–203.
21. Sanayei, A., Mousavi, S. F., & Yazdankhah, A. (2010). Group decision making process for supplier selection with VIKOR under fuzzy environment. *Expert Systems with Applications, 37*(1), 24–30.
22. Sayadi, M. K., Heydari, M., & Shahanaghi, K. (2009). Extension of VIKOR method for decision making problem with interval numbers. *Applied Mathematical Modelling, 33*(5), 2257–2262.
23. Shemshadi, A., Shirazi, H., Toreihi, M., & Tarokh, M. J. (2011). A fuzzy VIKOR method for supplier selection based on entropy measure for objective weighting. *Expert Systems with Applications, 38*(10), 12160–12167.
24. Tzeng, G. H., Teng, M. H., Chen, J. J., & Opricovic, S. (2002). Multicriteria selection for a restaurant location in Taipei. *International Journal of Hospitality Management, 21*(2), 171–187.
25. Wan, S. P., Wang, Q. Y., & Dong, J. Y. (2013). The extended VIKOR method for multi-attribute group decision making with triangular intuitionistic fuzzy numbers. *Knowledge-Based Systems, 52*, 65–77.
26. Yazdani, M., & Graeml, F. R. (2014). VIKOR and its applications: A state-of-the-art survey. *International Journal of Strategic Decision Sciences, 5*(2), 56–83.
27. Yu, P. L. (1973). A class of solutions for group decision problems. *Management Science, 19*(8), 936–946.
28. Yucenur, G. N., & Demirel, N. Ç. (2012). Group decision making process for insurance company selection problem with extended VIKOR method under fuzzy environment. *Expert Systems with Applications, 39*(3), 3702–3707.
29. Zeleny, M. (1982). *Multi criteria decision making*. New York: McGraw-Hills.
30. Zeng, Q. L., Li, D. D., & Yang, Y. B. (2013). VIKOR method with enhanced accuracy for multiple criteria decision making in healthcare management. *Journal of Medical Systems, 37*(2), 1–9.

Chapter 3
PROMETHEE

3.1 Introduction

The Preference Ranking Organization METHod for Enrichment of Evaluations (PROMETHEE) belongs to the outranking family of MCDA methods and was developed by Brans et al. [11, 13] and Brans and Vincke [12]. One of the creators of PROMETHEE, Professor Bertrand Mareschal,[1] maintains a full list of references to his website[2] that as of April 2017 numbered approximately 1,500 references, rendering the method to be quite popular. Input data is similar to TOPSIS and VIKOR, but the modeler is optionally required to feed the algorithm with a couple of more variables, depending on his preference function choice. The method has been later on complemented by GAIA (Geometrical Analysis for Interactive Aid), an attempt to represent the decision problem graphically in a two-dimensional plane. This visual interactive module can assist in complicated decision problems. GAIA, whilst interesting and of proven value to analysts, will not be presented here in detail as it requires a robust mathematical background from the reader. Instead this chapter will focus on the detailed analysis of the PROMETHEE methods and offer some insights on GAIA.

This is actually a family of methods, as PROMETHEE has been refined and extended over the years [10]. The main methodologies range from PROMETHEE I to VI and versions with interval, fuzzy numbers, and group decision making have

Electronic Supplementary Material The online version of this chapter (https://doi.org/10.1007/978-3-319-91648-4_3) contains supplementary material, which is available to authorized users.

[1]Professor Mareschal, in a discussion with the authors mentioned that for the names of PROMETHEE and GAIA some effort was required. The authors being Greek graciously acknowledge the effort! These are by no means the only MCDA methods bearing a Greek name - others like ORESTE and ELECTRE exist too.

[2]www.promethee-gaia.net.

J. Papathanasiou, N. Ploskas, *Multiple Criteria Decision Aid*, Springer Optimization and Its Applications 136, https://doi.org/10.1007/978-3-319-91648-4_3

Table 3.1 Distribution of papers on PROMETHEE by application areas [36][a]

Area	N	%
Environment management	323	21
Services and/or public applications	277	18
Industrial applications	243	15.8
Energy management	130	8.5
Water management	98	6.4
Finance	96	6.3
Transportation	59	3.8
Procurement	51	3.3
Other topics	68	4.4

[a]Some papers are related to multiple fields so that the total of the percentages is larger than 100%

also been developed [17–20, 25, 28, 29, 33, 34, 39, 42]. PROMETHEE results to a ranking of actions (as the alternatives are known in the method's terminology) and is based on preference degrees. Briefly, steps include the pairwise comparison of actions on each criterion, then the computation of unicriterion flows, and finally, the aggregation of the latter into global flows. It has been applied successfully in various application areas; Table 3.1 includes the applications of PROMETHEE according to the website of Professor Mareschal [36]. Application domains include nuclear waste management [14], productivity of agricultural regions [32], risk assessment [44], web site evaluation [5], renewable energy [16], environmental assessment [27], selection of contract type and project designer [3, 4].

3.2 Methodology

According to Brans and Mareschal [10], PROMETHEE is designed to tackle multicriteria problems such as the following

$$\max \{g_1(a), g_2(a), \cdots, g_n(a) | a \in A\} \tag{3.1}$$

where A is a finite set of possible alternatives $\{a_1, a_2, \cdots, a_m\}$ and $\{g_1(\cdot), g_2(\cdot), \cdots, g_n(\cdot)\}$ a set of evaluation criteria either to be maximized or minimized. The decision maker needs to construct the evaluation table as in Table 3.2. The second row of this table is about the weights associated with each of the criteria and as in the previous chapters, Equation (3.2) holds true.

$$\sum_{j=1}^{n} w_j = 1, \quad j = 1, 2, \cdots, n \tag{3.2}$$

Table 3.2 Evaluation table

a	$g_1(\cdot)$	$g_2(\cdot)$	$\cdots$	$g_n(\cdot)$
	w_1	w_2	$\cdots$	w_n
a_1	$g_1(a_1)$	$g_2(a_1)$	$\cdots$	$g_n(a_1)$
a_2	$g_1(a_2)$	$g_2(a_2)$	$\cdots$	$g_n(a_2)$
$\vdots$	$\vdots$	$\vdots$	$\ddots$	$\vdots$
a_m	$g_1(a_m)$	$g_2(a_m)$	$\cdots$	$g_n(a_m)$

It has to be pointed out that MCDA techniques in general place the decision makers in the center of the process and different decision makers can model the problem in different ways, according to their preferences (it also has to be mentioned here that the methods assist the decision maker, they do not make the final decision for him/her; thus, the word "aid" in the MCDA acronym. The responsibility for the final decision rests with the decision maker alone). In PROMETHEE, a preference degree is an expression of how one action is preferred against another action. For small deviations among the evaluations of a pair of criteria, the decision maker can allocate a small preference; if the deviation can be considered negligible, then this can be modeled in PROMETHEE too. The exact opposite stands for large deviations where the decision maker must allocate a large preference of one action over the other; if the deviation exceeds a certain value set by the decision maker, then there is an absolute preference of one action over the other. This preference degree is a real number always between 0 and 1.

Therefore the preference function, if the criterion is to be maximized, can be defined as

$$P_j(a,b) = F_j\left[d_j(a,b)\right], \quad \forall a, b \in A \tag{3.3}$$

where $d_j(a,b)$ is the difference of evaluations among two actions (pairwise comparison)

$$d_j(a,b) = g_j(a) - g_j(b) \tag{3.4}$$

and the preference degree is always between 0 and 1.

$$0 \leq P_j(a,b) \leq 1 \tag{3.5}$$

If criterion g has to be minimized, then $-g$ has to be maximized.

The pair $\left\{g_j(\cdot), P_j(a,b)\right\}$ is called by the authors of the method a generalized criterion associated to criterion $g_j(\cdot)$. They propose a total of six different types of preference functions as shown in Figure 3.1; these types have been accepted and used extensively in the literature. The rationale of the preference function is to model the way the decision maker prefers one action over another; it is actually an attempt to model his insights towards the problem in hand.

Generalized criterion		Definition	Parameters to fix
Type 1: Usual Criterion	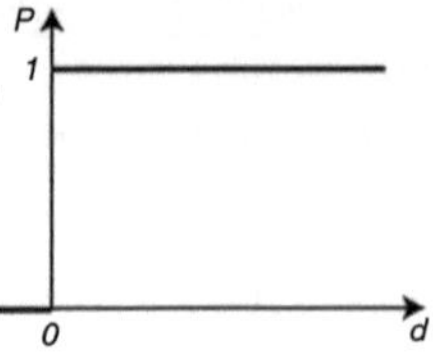	$P(d) = \begin{cases} 0 & d \le 0 \\ 1 & d > 0 \end{cases}$	–
Type 2: U-shape Criterion	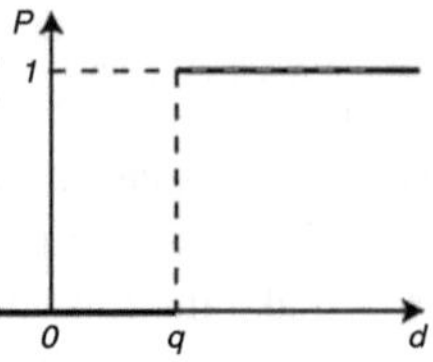	$P(d) = \begin{cases} 0 & d \le q \\ 1 & d > q \end{cases}$	q
Type 3: V-shape Criterion	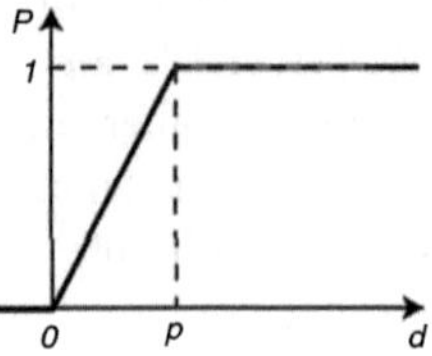	$P(d) = \begin{cases} 0 & d \le 0 \\ \frac{d}{p} & 0 \le d \le p \\ 1 & d > p \end{cases}$	p
Type 4: Level Criterion	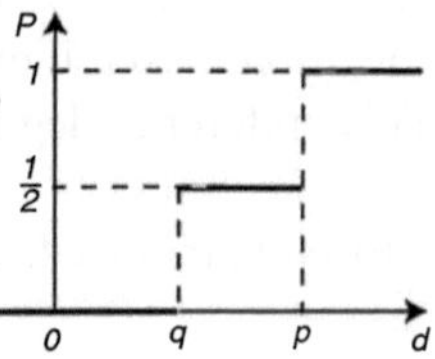	$P(d) = \begin{cases} 0 & d \le q \\ \frac{1}{2} & q < d \le p \\ 1 & d > p \end{cases}$	p, q
Type 5: V-shape with indifference Criterion	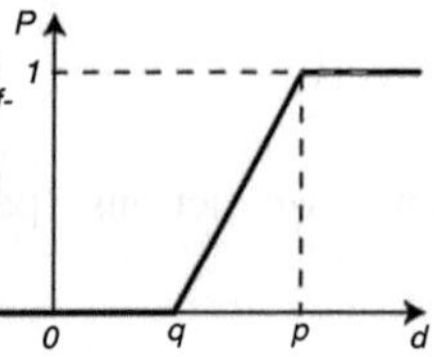	$P(d) = \begin{cases} 0 & d \le q \\ \frac{d-q}{p-q} & q < d \le p \\ 1 & d > p \end{cases}$	p, q
Type 6: Gaussian Criterion	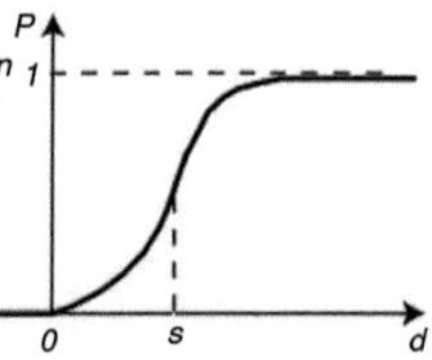	$P(d) = \begin{cases} 0 & d \le 0 \\ 1 - e^{-\frac{d^2}{2s^2}} & d > 0 \end{cases}$	s

Fig. 3.1 Types of generalized criteria [10]

In order to model whether a deviation (again, meaning the difference in the evaluations among a couple of actions in a single criterion) is negligible or denotes an absolute preference among a pair or actions, the decision maker has to input at most a couple of variables, q (the threshold of indifference, below which there is no preference to either of the actions meaning the preference degree is 0), p (the threshold of absolute preference, above which there is a total preference to one of the two actions and assigning the preference degree the value of 1), and in the case of a Gaussian function the inflection point s, which is an intermediate value between q and p. To quote Brans and Mareschal [10], "in case of a Gaussian criterion the preference function remains increasing for all deviations and has no discontinuities, neither in its shape, nor in its derivatives. A parameter s has to be selected; it defines the inflection point of the preference function. We then recommend to determine first q and p and to fix s in between. If s is close to q the preferences will be reinforced for small deviations, while close to p they will be softened." It is true that the introduction of the threshold fixing preference limits seems to add to the subjectivity of the method. However, it allows more flexibility in the modeling of the specific problem and in case the decision maker opts for the usual criterion no such parameters need to be fixed.

In more detail and in accordance with Figure 3.1, there are six types of preference functions:

1. Preference function type 1 (Usual criterion) requires no parameters q, p, or s to fix, indicating that the slightest deviation among a pair of actions is taken into account in the final ranking. The preference degree is either 0 (denoting an indifference between the actions, i.e., they are both considered equally preferred) if the deviation d among the pair of actions is 0, and 1 if the deviation is anything above 0 (denoting an absolute preference of one action against the other). This is the simplest case and useful when the decision maker cannot decide on the values of q and p. It makes also the method fully comparable with TOPSIS and VIKOR, as it does not require in this case the definition of additional variables (meaning q, p, or s).
2. Preference function type 2 (U-shape criterion) requires only parameter q to be fixed. This implies that the decision maker is able to fix a value for the indifference threshold, below which the actions are indifferent to him/her. However, the slightest value above the q parameter is crucial as the preference function instantly takes the value of 1. The preference function acts like a binary variable; it is either 0 or 1.
3. Preference function type 3 (V-shape criterion) requires only parameter p to be fixed. The preference function can take now any value between 0 and 1 ($P(d) = \frac{d}{p}, 0 \leq d \leq p$), but when the deviation is above the p threshold it acquires the value of 1.
4. Preference function type 4 (Level criterion) requires both p and q parameters to be fixed beforehand. Similar to the previous two cases, when the deviation is below q the preference degree is 0 and when the deviation is greater than p it equals to 1. Any value for the deviation between q and p means that the

preference degree is equal to $\frac{1}{2}$. Therefore, the preference function can only be either 0, $\frac{1}{2}$, or 1.

5. Preference function type 5 (V-shape with indifference criterion, also known as linear) is the same as type 4, only this time the preference degree does not have a fixed value between q and p, but it is a linear function acquiring any value between 0 and 1.
6. Preference function type 6 (Gaussian criterion) requires the decision maker to fix only parameter s. If the deviation is below or equal to 0, then the preference degree is 0; if the deviation is above 0, then the preference degree has the value of $1 - e^{-\frac{d^2}{2s^2}}$.

The aggregated preference indices can be calculated as follows:

$$\begin{cases} \pi(a,b) = \sum_{j=1}^{n} P_j(a,b)w_j \\ \pi(b,a) = \sum_{j=1}^{n} P_j(b,a)w_j \end{cases} \tag{3.6}$$

where $(a, b) \in A$, and $\pi(a, b)$ indicates how much action a is preferred to b over all of the criteria, while $\pi(b, a)$ how much action b is preferred to a. The following properties hold true for all $(a, b) \in A$

$$\begin{cases} \pi(a,a) = 0 \\ 0 \le \pi(a,b) \le 1 \\ 0 \le \pi(b,a) \le 1 \\ 0 \le \pi(a,b) + \pi(b,a) \le 1 \end{cases} \tag{3.7}$$

Each action is competing against $(m - 1)$ other actions in the set A. The unicriterion positive flow of each action in A is a number between 0 and 1 and is an indicator of how much this action is preferred over all the other actions in A. The higher this value is, the more this action is preferable for this particular decision maker. Therefore, the definition for the positive outranking flow is as

$$\phi^+(a) = \frac{1}{m-1} \sum_{x \in A} \pi(a,x) \tag{3.8}$$

On the other hand, the negative outranking flow is an indicator of how all the other actions are preferred over this particular action and in accordance with the positive flow is defined as

$$\phi^-(a) = \frac{1}{m-1} \sum_{x \in A} \pi(x,a) \tag{3.9}$$

The authors of PROMETHEE distinguish between PROMETHEE I and PROMETHEE II. The former provides the decision maker with a partial ranking of actions and the latter with a complete ranking; PROMETHEE II seems to have gained much more popularity in the literature than PROMETHEE I. PROMETHEE I needs both Φ^{+} and Φ^{-} rankings to produce the final ranking; when the two flows are conflicting, the particular actions are considered to be incomparable.

Therefore, and in more detail

$$\begin{cases} aP^{I}b \text{ iff } \begin{cases} \phi^{+}(a) > \phi^{+}(b) \text{ and } \phi^{-}(a) < \phi^{-}(b), \text{ or} \\ \phi^{+}(a) = \phi^{+}(b) \text{ and } \phi^{-}(a) < \phi^{-}(b), \text{ or} \\ \phi^{+}(a) > \phi^{+}(b) \text{ and } \phi^{-}(a) = \phi^{-}(b); \end{cases} \\ aI^{I}b \text{ iff } \quad \phi^{+}(a) = \phi^{+}(b) \text{ and } \phi^{-}(a) = \phi^{-}(b); \\ aR^{I}b \text{ iff } \begin{cases} \phi^{+}(a) > \phi^{+}(b) \text{ and } \phi^{-}(a) > \phi^{-}(b), \text{ or} \\ \phi^{+}(a) < \phi^{+}(b) \text{ and } \phi^{-}(a) < \phi^{-}(b); \end{cases} \end{cases} \tag{3.10}$$

where P^{I}, I^{I}, and R^{I} stand for preference, indifference, and incomparability, respectively (the superscript I stands for PROMETHEE I).

PROMETHEE II, on the other hand, is based on the net flow Φ, which is defined as

$$\Phi(a) = \Phi^{+}(a) - \Phi^{-}(a) \tag{3.11}$$

The action is better as the net flow value is larger.

$$\begin{cases} aP^{II}b \text{ iff } \phi(a) > \phi(b) \\ aI^{II}b \text{ iff } \phi(a) = \phi(b) \end{cases} \tag{3.12}$$

As mentioned earlier, PROMETHEE II results in a complete ranking of actions and there are not any incomparabilities; the code presented later in this chapter will be for this version. However, the reader can easily modify it to get PROMETHEE I results. Although PROMETHEE II is much more frequently used over PROMETHEE I, the authors (Brans and Mareschal [11]) state that "In real-world applications, we recommend to both the analysts and the decision makers to consider both PROMETHEE I and PROMETHEE II. The complete ranking is easy to use, but the analysis of the incomparabilities often helps to finalize a proper decision."

According to Equations (3.6), (3.8), (3.9), and (3.11), the net flow is as

$$\phi(a) = \phi^{+}(a) - \phi^{-}(a) = \frac{1}{m-1} \sum_{j=1}^{n} \sum_{x \in A} \left[P_j(a,x) - P_j(x,a)\right] w_j \tag{3.13}$$

and

$$\phi(a) = \sum_{j=1}^{n} \phi_j(a) w_j \qquad (3.14)$$

where the weights w_j, $j = 1, 2, \cdots, n$, are also considered.

Ishizaka and Nemery in their excellent volume about MCDA [30] argue that there are two different ways of computing the global flows; both producing the same results. Figure 3.2 is adopted from their work and presents the steps of the method. In the code presented later in this chapter, the flow on the left side was implemented. However, the Python code can be easily modified to produce the global positive and negative flows as well.

Considering all of the above, the steps of the PROMETHEE algorithm can be summarized as follows:

1. Build the decision matrix and set the values of the preference parameters and the weights.
2. Calculate the deviations between the evaluations of the alternatives in each criterion.
3. Calculate the pairwise comparison matrix for each criterion.
4. Calculate the unicriterion net flows.
5. Calculate the weighted unicriterion flows.
6. Calculate the global preference net flows.
7. Rank the actions according to PROMETHEE I or PROMETHEE II rules.

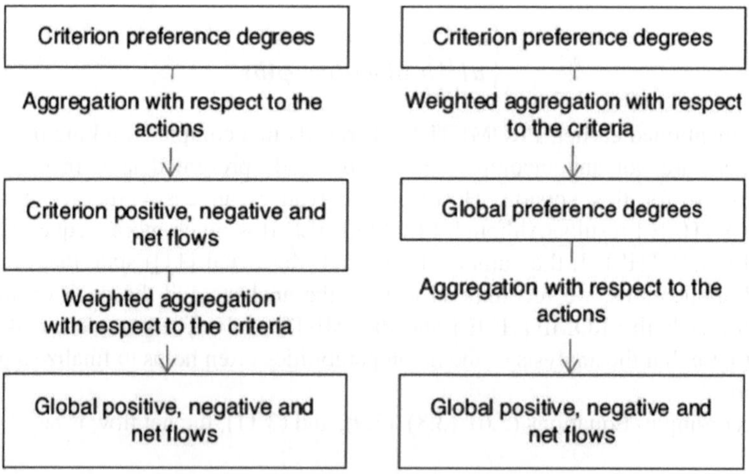

Fig. 3.2 Alternative ways of computing the global flows based on criterion preference degrees [30]

There are not many software packages available for PROMETHEE today. During the 80s the method's original authors developed PROMCALC [8]; a software package that runs under MSDOS. It later evolved to Decision Lab 2000 with a visual interface and full documentation. This tool is no longer supported. Therefore, three other alternatives exist: Smart Picker Pro and DECERNS [43], both of which are proprietary software and require a license. Luckily enough Visual PROMETHEE exists, developed by Professor Bertrand Mareschal, who provides an academic edition free of charge.

As a final note to this section, the family also includes PROMETHEE III, IV, and VI. Those family members have not been used much; there are only few papers using these methods, like Cavalcante and de Almeida [15], Kaya et al. [31], and Albuquerque [2]. Finally, the rank reversal phenomenon has been studied in PROMETHEE as well [23, 37].

3.2.1 Numerical Example

To allow the reader to compare PROMETHEE with TOPSIS and VIKOR presented in previous chapters, the same example will be used. Table 1.2 is repeated here as Table 3.3 extended with the preference function choice and relevant data. Conveniently enough, Oprikovic and Tzeng [40] published a paper actually comparing in depth these three methods; the reader is advised to refer on that work for more details.

The second step of PROMETHEE involves the calculation of the deviations between the evaluations of the alternatives in each criterion. In this example, and for the criterion investment costs, the differences between the evaluations of the actions (sites) are in Table 3.4; one such table is required for each of the criteria. Note that the main diagonal of the table has all elements equal to zero, since one action cannot ever be preferred over itself.

Table 3.3 Initial data for the numerical example (extended Table 1.2)

	Investment costs (million €)	Employment needs (hundred employees)	Social impact (1–7)	Environmental impact (1–7)
Weight	0.4	0.4	0.1	0.2
Preference function	Linear ($q = 1$, $p = 2$)	Linear ($q = 1$, $p = 2$)	Linear ($q = 1$, $p = 2$)	Linear ($q = 1$, $p = 2$)
Site 1	8	7	2	1
Site 2	5	3	7	5
Site 3	7	5	6	4
Site 4	9	9	7	3
Site 5	11	10	3	7
Site 6	6	9	5	4

Table 3.4 Differences between the evaluations of the sites on the investment costs criterion

	Site 1	Site 2	Site 3	Site 4	Site 5	Site 6
Site 1	0	3	1	−1	−3	2
Site 2	−3	0	−2	−4	−6	−1
Site 3	−1	2	0	−2	−4	1
Site 4	1	4	2	0	−2	3
Site 5	3	6	4	2	0	5
Site 6	−2	1	−1	−3	−5	0

Next, we calculate the pairwise comparison matrix for each criterion. In the investment criterion again, which is supposed to be maximized, site 2 compared to site 1 yields a value of −3 ($d = 5 - 8 = -3$, as shown in Table 3.4). Depending on the preference function choice, the following cases can occur (Figure 3.1):

1. Preference function type 1 (Usual criterion): Parameters p and q are not fixed and $d \leq 0$. Therefore, $P(d) = 0$; there is no preference of site 2 over site 1.
2. Preference function type 2 (U-shape criterion): Parameter q should be fixed; since $d = -3$, then regardless of the value of q, $P(d) = 0$ in all cases.
3. Preference function type 3 (V-shape criterion): Parameter p needs to be predefined, and the case is the same as the previous one.
4. Preference function type 4 (Level criterion): In a possible scenario, the decision maker opts for the values 1 and 3 for q and p, respectively (or any other values he/she thinks are best). The case remains the same as the previous one.
5. Preference function type 5 (V-shape with indifference criterion). Again, same as the previous case.
6. Preference function type 6 (Gaussian criterion): Only parameter s needs to be fixed. However, the case is the same as before since $d = -3$.

In all of the above cases the preference of site 2 over site 1 is 0. However, more interesting is to check how site 1 compares to site 2 in the same criterion. The deviation among the evaluations is now 3 and the following cases may occur:

1. Preference function type 1 (Usual criterion): Parameters p and q are not fixed and $d \geq 0$. Therefore, $P(d) = 1$; there is a strong (absolute) preference of site 1 over site 2.
2. Preference function type 2 (U-shape criterion): Parameter q should be fixed; since $d = 3$, then if q takes any value less than 3, $P(d) = 1$; $P(d) = 0$ in all other cases.
3. Preference function type 3 (V-shape criterion): Parameter p needs to be predefined. If p is greater than or equal to d ($p \geq 3$), then $P(d) = \frac{d}{p}$, else if $p < 3$ then $P(d) = 1$. Let's consider for this example that the decision maker fixes p at a value of 2; consequently $P(d) = 1$.
4. Preference function type 4 (Level criterion): In this scenario, the decision maker inputs the values 1 and 2 for q and p, respectively. Then P(d) is set to 1 (as previously, since $d > p$).

5. Preference function type 5 (V-shape with indifference criterion): Same as the previous case for $q = 1$ and $p = 2$.
6. Preference function type 6 (Gaussian criterion): For $s = 3$, $P(d) = 0.39$.

Finally, Table 3.5 is produced considering a linear preference function for the criterion investment costs (where $q = 1$ and $p = 2$).

One such table needs to be calculated for each criterion.

The fourth step of PROMETHEE involves the calculation of the unicriterion net flows. To obtain the value of the positive outranking flow for site 1 in this example, the decision maker has to sum the values of the first line of Table 3.5 (the main diagonal element is naturally 0) and divide the result with the number of actions minus 1, since a site is not compared with itself. In this case, the positive flow for site 1 is calculated as $\frac{1+0+0+0+1}{5}$ and equals to 0.4 (always for the investment criterion). For the value of the negative outranking flow of site 1, the decision maker has to sum all the elements of column 1 of Table 3.5 (again the main diagonal element equals to 0) and divide the result with the number of actions minus 1. In this case, the negative flow for site 1 is $\frac{0+0+0+1+0}{5} = 0.2$. From this point and on, the net flow is easily calculated as the difference between the positive and the negative flow; for site 1 being $0.4 - 0.2 = 0.2$. Bearing in mind the values that the positive and negative flows can obtain, the net score always lies between -1 and 1. For all actions in this example, the positive, negative, and net flows are in Table 3.6 for the investment criterion.

So far only the investment criterion has been taken into account. If the same procedure is applied to all other criteria and choosing a linear preference function (with $q = 1$ and $p = 2$, as in Table 3.3) in all cases, then Table 3.7 can be produced showing the net flows for all criteria.

Table 3.5 Pairwise comparison matrix for the criterion investment costs

	Site 1	Site 2	Site 3	Site 4	Site 5	Site 6
Site 1	0	1	0	0	0	1
Site 2	0	0	0	0	0	0
Site 3	0	1	0	0	0	0
Site 4	0	1	1	0	0	1
Site 5	1	1	1	1	0	1
Site 6	0	0	0	0	0	0

Table 3.6 Positive, negative, and net flows for the investment criterion

Action	Positive flow (Φ^+)	Negative flow (Φ^-)	Net flow (Φ)
Site 1	0.4	0.2	0.2
Site 2	0	0.8	−0.8
Site 3	0.2	0.4	−0.2
Site 4	0.6	0.2	0.4
Site 5	1	0	1
Site 6	0	0.6	−0.6

Next, the weighted unicriterion net flows are produced by multiplying each column of Table 3.7 by the corresponding weight associated with each column (the weights of the criteria are shown in Table 3.3); the results are shown in Table 3.8.

The sixth step of PROMETHEE involves the calculation of the global preference net flows. In Table 3.9, the second column (net flow) can be produced by summing up the (weighted) unicriterion flows in each row of Table 3.8; in the case of site 1 that means $0.08 + (-0.06) + (-0.08) + (-0.02) = -0.26$.

Finally, we rank the actions according to PROMETHEE I or PROMETHEE II rules. As stated earlier, when the two flows are conflicting, the particular actions are considered to be incomparable and the ranking is partial; this is the case between sites 6 and 1 and sites 1 and 3 in Figure 3.3 (their respective lines intersect). The best action is site 5.

Table 3.7 Unicriterion preference net flows

	Investment costs	Employment needs	Social impact	Environmental impact
Site 1	0.2	−0.2	−0.8	−1
Site 2	−0.8	−1	0.6	0.2
Site 3	−0.2	−0.6	0.4	0
Site 4	0.4	0.6	0.6	−0.2
Site 5	1	0.6	−0.8	1
Site 6	−0.6	0.6	0	0

Table 3.8 Weighted unicriterion preference net flows

	Investment costs	Employment needs	Social Impact	Environmental impact
Site 1	0.08	−0.06	−0.08	−0.2
Site 2	−0.32	−0.3	0.06	0.04
Site 3	−0.08	−0.18	0.04	0
Site 4	0.16	0.18	0.06	−0.04
Site 5	0.4	0.18	−0.08	0.2
Site 6	−0.24	0.18	0	0

Table 3.9 Global preference net flows

	Net flow (Φ)
Site 1	−0.26
Site 2	−0.52
Site 3	−0.22
Site 4	0.36
Site 5	0.7
Site 6	−0.06

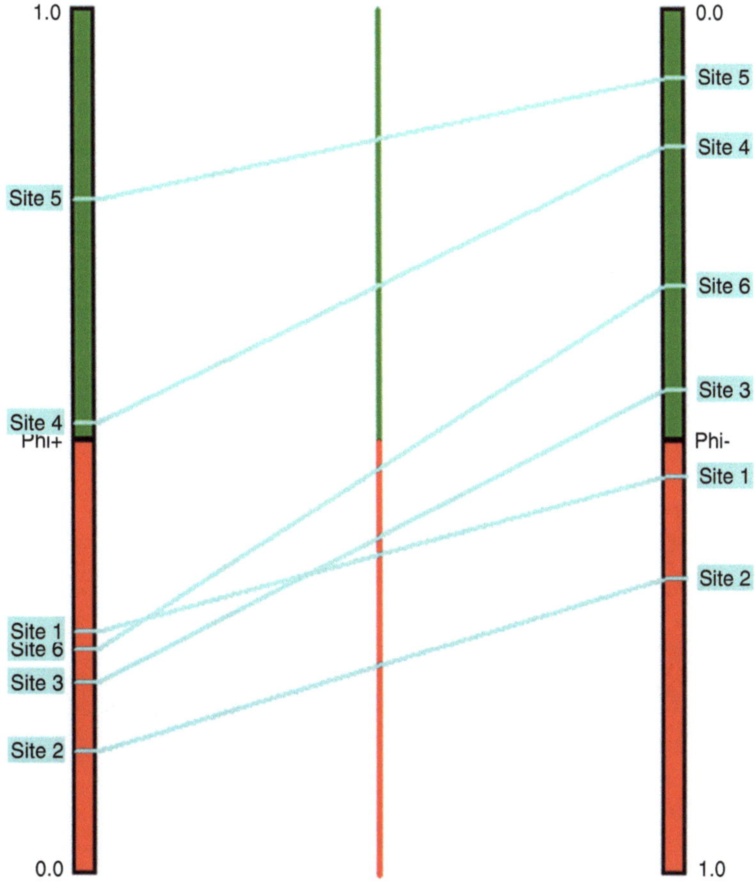

Fig. 3.3 PROMETHEE I ranking (figure produced by Visual PROMETHEE)

Figure 3.4 presents the results of PROMETHEE II using the Visual PROMETHEE software. It is clearly evident that site 5 is the best action and this time the ranking is complete.

As already mentioned, there are two different ways of computing the global flows (Figure 3.2). If the right side of Figure 3.2 is used, then Table 3.10 is produced with the results including the global positive and negative flows. This table is produced using the Visual PROMETHEE software; however, the Python code presented later in this chapter can be easily modified to produce the global positive and negative flows as well.

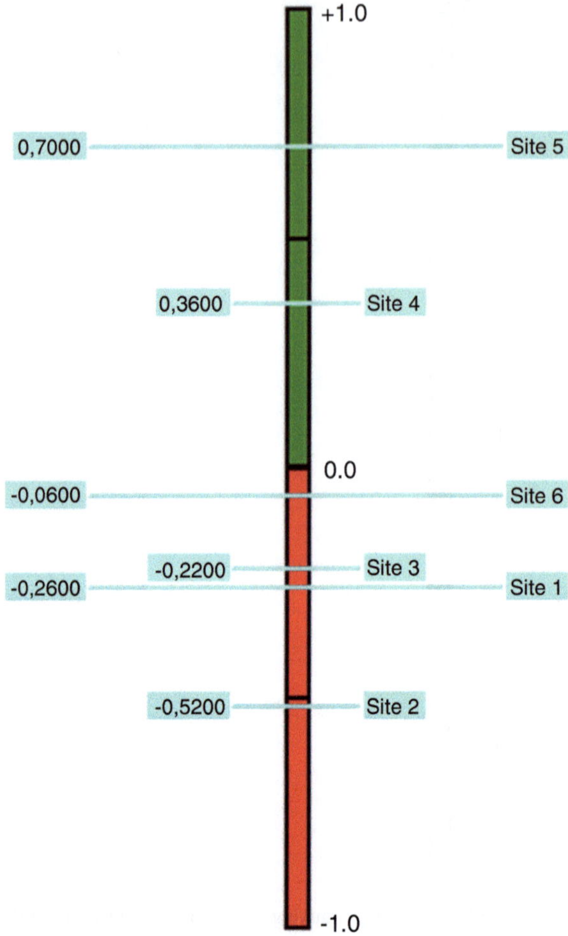

Fig. 3.4 PROMETHEE II ranking (figure produced by Visual PROMETHEE)

Table 3.10 Global preference flows

	Positive flow (Φ^+)	Negative flows (Φ^-)	Net flows (Φ)
Site 1	0.28	0.54	−0.26
Site 2	0.14	0.66	−0.52
Site 3	0.22	0.44	−0.22
Site 4	0.52	0.16	0.36
Site 5	0.78	0.08	0.7
Site 6	0.26	0.32	−0.06

3.2.2 Python Implementation

The implementation of PROMETHEE in Python is organized in two main modules and an optional one. The file *PROMETHEE_II.py* includes a Python implementation of PROMETHEE II. The input variables are arrays *x* (action performances, lines 45–46), *p* (preference parameters of all criteria, line 49), *c* (criteria optimization, line 52), *d* (preference functions, line 55), and *w* (weights of criteria, line 58). The function *promethee* implements PROMETHEE II method calling all other functions (lines 11–35). The final results are displayed in lines 62–66.

```
1.  # Filename: PROMETHEE_II.py
2.  # Description: PROMETHEE II method
3.  # Authors: Papathanasiou, J. & Ploskas, N.
4.
5.  import matplotlib.pyplot as plt
6.  from numpy import *
7.  from PROMETHEE_Preference_Functions import uni_cal
8.  from PROMETHEE_Final_Rank_Figure import graph, plot
9.
10. # PROMETHEE method: it calls the other functions
11. def promethee(x, p, c, d, w):
12.     """ x is the action performances array, b is the
13.     array with the preference parameters of all
14.     criteria, c is the criteria min (0) or max (1)
15.     optimization array, d is the preference
16.     function array ('u' for usual, 'us' for
17.     u-shape, 'vs' for v-shape, 'le' for level,
18.     'li' for linear, and 'g' for Gaussian), and w
19.     is the weights array
20.     """
21.     weighted_uni_net_flows = []
22.     total_net_flows = []
23.     for i in range(x.shape[1]):
24.         weighted_uni_net_flows.append(w[i] *
25.             uni_cal(x[:, i:i + 1], p[:,
26.             i:i + 1], c[i], d[i]))
27.
28.     # print the weighted unicriterion preference
29.     # net flows
30.     for i in range(size(weighted_uni_net_flows, 1)):
31.         k = 0
32.         for j in range(size(weighted_uni_net_flows, 0)):
33.             k = k + round(weighted_uni_net_flows[j][i], 5)
34.         total_net_flows.append(k)
35.     return around(total_net_flows, decimals = 4)
36.
37. # main function
38. def main(a, b):
39.     """ a and b are flags; if they are set to 'y' they do
40.     print the results, anything else does not print
41.     the results
42.     """
```

```

    # action performances array
    x = array([[8, 7, 2, 1], [5, 3, 7, 5], [7, 5, 6, 4],
        [9, 9, 7, 3], [11, 10, 3, 7], [6, 9, 5, 4]])

    # preference parameters of all criteria array
    p = array([[1, 1, 1, 1], [2, 2, 2, 2]])

    # criteria min (0) or max (1) optimization array
    c = ([1, 1, 1, 1])

    # preference function array
    d = (['li', 'li', 'li', 'li'])

    # weights of criteria
    w = array([0.4, 0.3, 0.1, 0.2])

    # final results
    final_net_flows = promethee(x, p, c, d, w)
    print("Global preference flows = ", final_net_flows)
    if a == 'y':
        graph(final_net_flows, "Phi")
    if b == 'y':
        plot(final_net_flows, "PROMETHEE II")
    return final_net_flows

if __name__ == '__main__':
    main('n','y')
```

The file *PROMETHEE_Preference_Functions.py* includes a method that calculates the unicriterion preference degrees of the actions for a specific criterion. Method *uni_cal* takes as input arrays *x* (action performances), *p* (preference parameters of all criteria), *c* (criteria optimization), and *f* (preference functions), and returns as output the net flows; positive and negative flows are also computed (these will be needed if the reader decides to modify the code for PROMETHEE I).

```
# Filename: PROMETHEE_Preference_Functions.py
# Description: This module calculates the
# unicriterion preference degrees of the actions
# for a specific criterion
# Authors: Papathanasiou, J. & Ploskas, N.

from numpy import *

# Calculate the unicriterion preference degrees
def uni_cal(x, p, c, f):
    """ x is the action performances array, p is the
    array with the preference parameters of all
    criteria, c is the criteria min (0) or max (1)
    optimization array, and f is the preference
    function array for a specific criterion ('u'
    for usual, 'us' for u-shape, 'vs' for v-shape,
    'le' for level, 'li' for linear, and 'g' for
```

```
    Gaussian)
    """
    uni = zeros((x.shape[0], x.shape[0]))
    for i in range(size(uni, 0)):
        for j in range(size(uni, 1)):
            if i == j:
                uni[i, j] = 0
            elif f == 'u':  # Usual preference function
                if x[j] - x[i] > 0:
                    uni[i, j] = 1
                else:
                    uni[i, j] = 0
            elif f == 'us': # U-shape preference function
                if x[j] - x[i] > x[0]:
                    uni[i, j] = 1
                elif x[j] - x[i] <= p[0]:
                    uni[i, j] = 0
            elif f == 'vs': # V-shape preference function
                if x[j] - x[i] > p[1]:
                    uni[i, j] = 1
                elif x[j] - x[i] <= 0:
                    uni[i, j] = 0
                else:
                    uni[i, j] = (x[j] - x[i]) / p[1]
            elif f == 'le': # Level preference function
                if x[j] - x[i] > p[1]:
                    uni[i, j] = 1
                elif x[j] - x[i] <= p[0]:
                    uni[i, j] = 0
                else:
                    uni[i, j] = 0.5
            elif f == 'li': # Linear preference function
                if x[j] - x[i] > p[1]:
                    uni[i, j] = 1
                elif x[j] - x[i] <= p[0]:
                    uni[i, j] = 0
                else:
                    uni[i, j] = ((x[j] - x[i]) -
                        p[0]) / (p[1] - p[0])
            elif f == 'g':  # Gaussian preference function
                if x[j] - x[i] > 0:
                    uni[i, j] = 1 - math.exp(-(math.pow(x[j]
                        - x[i], 2) / (2 * p[1] ** 2)))
                else:
                    uni[i, j] = 0
    if c == 0:
        uni = uni
    elif c == 1:
        uni = uni.T
    # positive, negative and net flows
    pos_flows = sum(uni, 1) / (uni.shape[0] - 1)
    neg_flows = sum(uni, 0) / (uni.shape[0] - 1)
    net_flows = pos_flows - neg_flows
    return net_flows
```

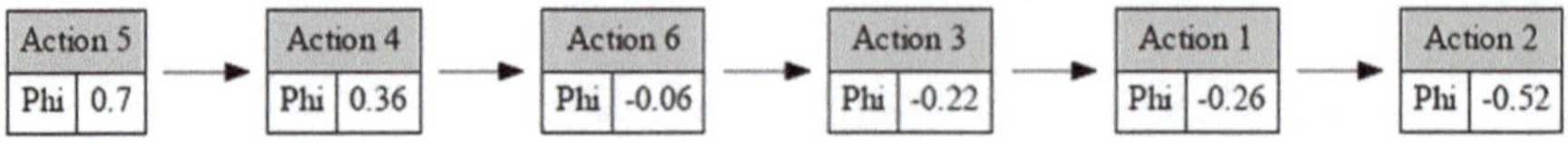

Fig. 3.5 Final rank figure plotted with graphviz module

Fig. 3.6 Final rank plotted with matplotlib

The file *PROMETHEE_Final_Rank_Figure.py* is an optional module that includes two methods to plot the results of PROMETHEE. This module produces two plots with the results (Figures 3.5 and 3.6). In order to run these methods, one needs to install graphviz and matplotlib modules.

```
# Filename: PROMETHEE_Final_Rank_Figure.py
# Description: Optional module to plot the
# results of PROMETHEE method
# Authors: Papathanasiou, J. & Ploskas, N.

import matplotlib.pyplot as plt
from graphviz import Digraph
from numpy import *

# Plot final rank figure
def graph(flows, b):
    """ flows is the matrix with the net flows, and b
```

```
    is a string describing the net flow
    """
    s = Digraph('Actions', node_attr = {'shape':
        'plaintext'})
    s.body.extend(['rankdir = LR'])
    x = sort(flows)
    y = argsort(flows)
    l = []
    for i in y:
        s.node('action' + str(i), '''<
        <TABLE BORDER="0" CELLBORDER="1"
            CELLSPACING="0" CELLPADDING="4">
          <TR>
            <TD COLSPAN="2" bgcolor="grey" >Action
                ''' + str(y[i] + 1) + '''</TD>
          </TR>
          <TR>
            <TD>'''+ b +'''</TD>
            <TD>''' + str(x[i]) + '''</TD>
          </TR>
        </TABLE>>''')
    k = []
    for q in range(len(flows) - 1):
        k.append(['action' + str(q + 1), 'action'
            + str(q)])
    print(k)
    s.edges(k)
    s.view()

# Plot final rank
def plot(a, b):
    """ a is the matrix with the net flows, and b
    is a string describing the method
    """
    flows = a
    yaxes_list = [0.2] * size(flows, 0)
    plt.plot(yaxes_list, flows, 'ro')
    frame1 = plt.gca()
    frame1.axes.get_xaxis().set_visible(False)
    plt.axis([0, 0.7, min(flows) - 0.05,
        max(flows) + 0.05])
    plt.title(b + " results")
    plt.ylabel("Flows")
    plt.legend()
    plt.grid(True)
    z1 = []
    for i in range(size(flows, 0)):
        z1.append('   (Action ' + str(i + 1) + ')')
    z = [str(a) + b for a, b in zip(flows, z1)]
    for X, Y, Z in zip(yaxes_list, flows, z):
        plt.annotate('{}'.format(Z), xy = (X, Y),
            xytext=(10, -4), ha = 'left',
            textcoords = 'offset points')
    plt.show()
```

The preference function choice is important; it is a typical procedure during the decision making process for the modeler to test various scenarios and compare the results. If the first criterion is to be minimized (all the other criteria are maximized), then Table 3.11 includes a couple of different scenarios. In the first scenario, the decision maker is unable to define with adequate accuracy the q and p values, so he/she opts for the 'usual' preference function. In the second, after some effort, he/she managed to obtain some of these values and used alternative preference functions in each criterion. Table 3.12 has the global net flows for each of the scenarios and Figure 3.7 displays the ranking in each case. It is clear that the ranking is not the same and this is a problem with small deviations among the input data numbers; in larger problems, the ranking may change considerably.

Table 3.11 Alternative scenarios

	Criterion	Preference function	Q (indifference threshold)	P (absolute preference threshold)
Scenario 1	Investment costs	Usual	–	–
	Employment needs	Usual	–	–
	Social impact	Usual	–	–
	Environmental impact	Usual	–	–
Scenario 2	Investment costs	Usual	–	–
	Employment needs	U-shape	1	–
	Social impact	V-shape	–	2
	Environmental impact	Linear	1	2

Table 3.12 Scenario rankings (PROMETHEE II)

	Scenario 1 net flow (Φ)	Scenario 2 net flow (Φ)
Site 1	−0.44	−0.43
Site 2	0.3	0.21
Site 3	−0.08	−0.07
Site 4	−0.16	−0.03
Site 5	0.04	−0.09
Site 6	0.34	0.41

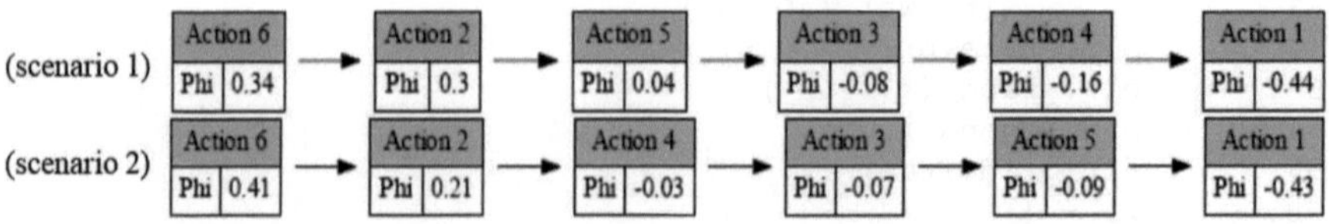

Fig. 3.7 Final ranking figure comparison of the different scenarios

3.2.3 *Visual Decision Aid: The GAIA Method*

The GAIA visual module is based on the principal component analysis technique [8, 9]; it provides the decision maker with a visual representation of the Φ matrix ($\phi_j(a_i)$, where $i = 1, 2, \cdots, m$ and $j = 1, 2, \cdots, n$) in space $\mathbb{R}^n$. It is an attempt to visualize in two dimensions a decision problem. Using the Visual PROMETHEE software, the image produced for the site selection example (Table 3.3) is displayed in Figure 3.8. To correctly interpret the image, the decision maker should keep in mind that the following rules apply:

- the criteria are represented by axes. The four criteria of the decision problem in hand are the four axes ending in a diamond. The codenames of the criteria are C_1 to C_4.
- the length of each of the axes is important; this depends upon the deviations among the evaluations of the actions in a criterion. The longer the axis, the larger the deviations of the net flow values. The environmental impact criterion (C_4) in

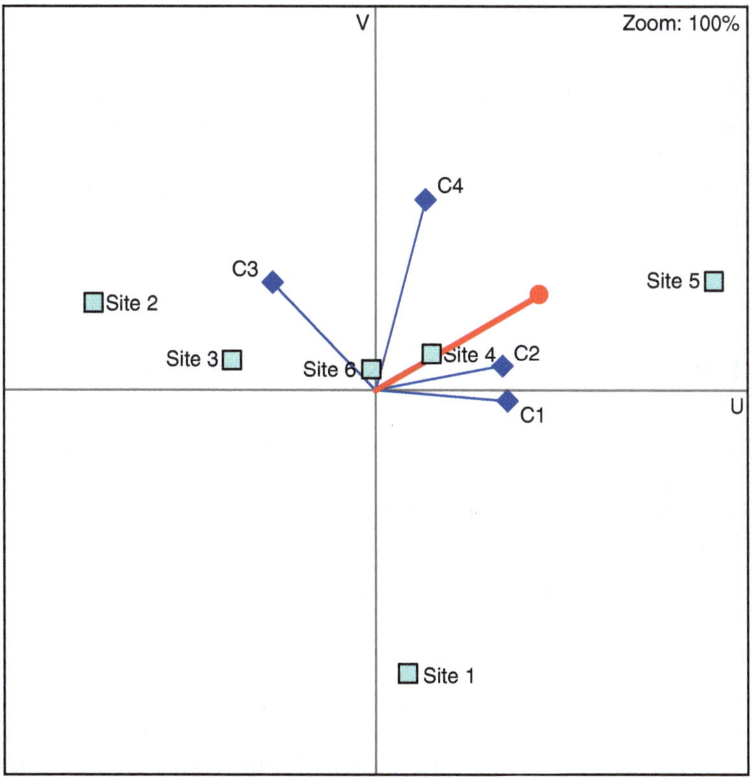

Fig. 3.8 GAIA plane for the site selection problem (figure produced by Visual PROMETHEE)

this case has larger deviations among its action evaluations than the rest of the criteria; thus, the longer axes.

- the orientation of the criteria axis is also important. It serves as an indication of the 'agreement' or conflict between the criteria. If the angle between a couple of axes is rather small, the corresponding criteria are 'in agreement' or are correlated. In the case that the angle is large, the criteria are conflicting among them; this represents a problem that the decision maker must take into account. Investment costs (C_1) and employment needs (C_2) are well correlated, but both are in conflict with the social impact (C_3) criterion.
- actions are represented by squares; their relative positions imply whether their profiles are similar or not. The closer two squares are, the more similar the actions; the further means that the actions are very different. Sites 4 and 6 are very similar whilst site 1 is very different from all the other sites. Note that the indifference and preference thresholds are important in order to define the concept of similarity.
- the criteria weights information is also present in the image. The thick line ending on a circle is called the decision stick. When the weights change the actions and the axes remain in the same positions, what changes is the decision stick. The projections of the actions in this stick show their relative positions in the PROMETHEE II ranking.
- finally, some amount of information is lost when representing the decision problem data in two dimensions. GAIA (and Visual PROMETHEE) calculates the retained information in the GAIA plane; if the value quality is above 70%, the plane is considered reliable. In this example, the preserved information is 76.6%, meaning the representation has maintained enough information through the transformations.

3.3 PROMETHEE Group Decision Support System

To tackle group decision making problems, the PROMETHEE GDSS (Group Decision Support System) method has been developed. Brans et al. [6], and Brans and Mareschal [10] propose a three-phase procedure that includes 11 steps. The three phases are:

1. Phase 1: Generation of alternatives and criteria.
2. Phase 2: Individual evaluation by each decision maker.
3. Phase 3: Global evaluation by the group.

Let's assume that the decision to locate a new facility is a group decision making problem and that three different stakeholders, each with his/her own agenda, are engaged in the procedure. After some discussions, they agree on the criteria and the alternatives, but maintain some differences in the evaluations, especially for the social impact and environmental impact criteria that are qualitative and not quantitative. Furthermore, they have to agree on the preference functions used and the weights of the criteria. Instead of using methodologies like Goal Programming and Simos (or revised Simos) method to obtain the weights (and this can easily prove to be a prolonged and tedious task in a group decision making environment with a lot of competitive domain experts supporting different views), PROMETHEE GDSS allows each decision maker to keep his/her own evaluations of the weights and the preference functions. Once all involved stakeholders have acquired the final ranking with PROMETHEE II, then a global matrix is constructed where the alternatives are again the rows and the columns include the decision makers net flow values from the individual rankings. In this global matrix, the weights can be equal for all involved parties or if they agreed to differ (that situation can occur for various reasons; funding level by each party, problem awareness and knowledge level for the experts, etc.), the weights may not be equal. The preference functions can be of the usual type, if the decision makers cannot agree on the q and p values or any other type if they can reach a consensus on this point. When the global matrix is constructed, the final ranking is produced; Figure 3.9 describes the whole procedure for three

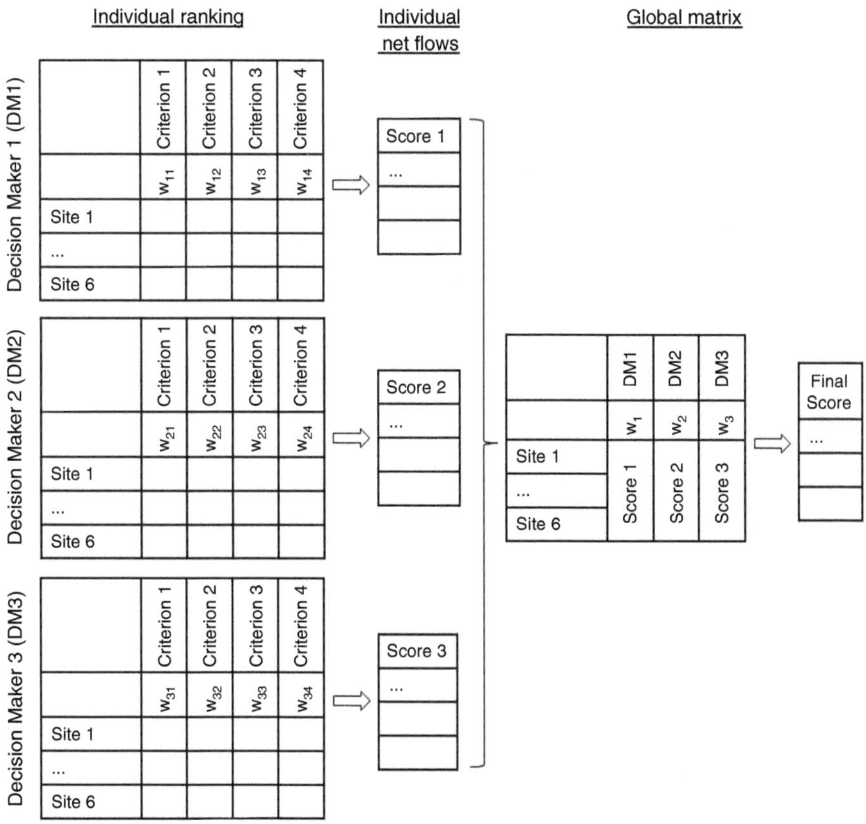

Fig. 3.9 The PROMETHEE GDSS procedure

decision makers, but the procedure is similar for any other number of decision makers. The readers can easily extend and modify the Python code provided in this chapter as necessary to address the group decision making problem.

3.4 PROMETHEE V

In some cases, the decision maker does not need to rank the actions from best to worst, but to choose a subset (a 'portfolio') from the actions, under a number of constraints. This is the case where PROMETHEE V comes in focus. Mavrotas et al. [38], De Almeida and Vetchera [21], and De Almeida et al. [22] have argued that the method has some issues of scale. De Almeida and Vetchera [21] proposed improvements based on the concept of a c-optimal portfolio. However, in this section we present the original method as proposed by Brans and Mareschal [7, 10]. PROMETHEE V has been used in project portfolio selection [35], water supply management [1, 26], capital budgeting [24] among other application domain. A linear programming solver must be used in order to solve such problems. There are many linear programming solvers available and most of them can be used in Python. CLP[3] is a good open source alternative that can tackle large scale linear programming problems. In addition, CPLEX[4] is a powerful commercial linear programming solver, free for academic use.

Supposing the decision maker has concluded on a set of possible actions, a_i, where $i = 1, 2, \cdots, m$, a boolean variable is associated to each one of them

$$x_i = \begin{cases} 1 \text{ if } a_i \text{ is selected} \\ 0 \quad\quad \text{if not} \end{cases}$$

Then, the procedure goes on with the following steps:

1. Initially, the multicriteria problem is considered without any constraints. The net flows, $\phi(a_i)$, where $i = 1, 2, \cdots, m$, are computed according to the PROMETHEE II ranking procedure.
2. A binary linear model is then formulated as follows:

$$\begin{aligned} \min z = & \quad \textstyle\sum_{i=1}^{n} \phi(a_i)x_i \\ \text{s.t.} & \quad \textstyle\sum_{i=1}^{m} \lambda_{p,i}x_i \sim \beta_p, \; p = 1, 2, \cdots, P \\ x_i \in & \{0, 1\}, i = 1, 2, \ldots, m \end{aligned}$$

where $\sim$ stands for $=, \geq$, or $\leq$.

[3]https://projects.coin-or.org/Clp.

[4]https://www-01.ibm.com/software/commerce/optimization/cplex-optimizer/.

The above model aims in selecting the actions that have the largest possible net flow, taking also into account the constraints.

In this chapter, we are using PuLP [41] to solve linear programming problems. PuLP is an open source Python-based linear programming solver that allows the user to optimize mathematical programming models. It supports both linear and mixed integer programming and maintains a simple and intuitive syntax with an easy learning curve. This is not the only package available in Python for solving linear programming problems; Pyomo, which is presented and used in Chapter 6, is also available. Yet the learning curve in the latter case is much more steeper and demanding (however, Pyomo is more powerful than PuLP). PuLP supports a number of solvers, in this section the included default solver will be used; CLP or CPLEX can be utilized as well.

We will demonstrate the basic usage of PuLP by solving two simple examples. Initially, we will solve the following simple linear programming problem

$$\begin{aligned} \min z = {} & 60x_1 + 40x_2 \\ \text{s.t.} \quad & 4x_1 + 4x_2 \geq 10 \\ & 2x_1 + x_2 \geq 4 \\ & 6x_1 + 2x_2 \leq 12 \\ & x_1 \geq 0, x_2 \geq 0 \end{aligned}$$

The file *PULP_example_1.py* includes a Python implementation for solving this linear programming problem with PuLP. Since it has only a couple of decision variables the results can be plotted as shown in Figure 3.10. This figure includes an abundance of information; the feasible region (the grey shaded area), the constraints

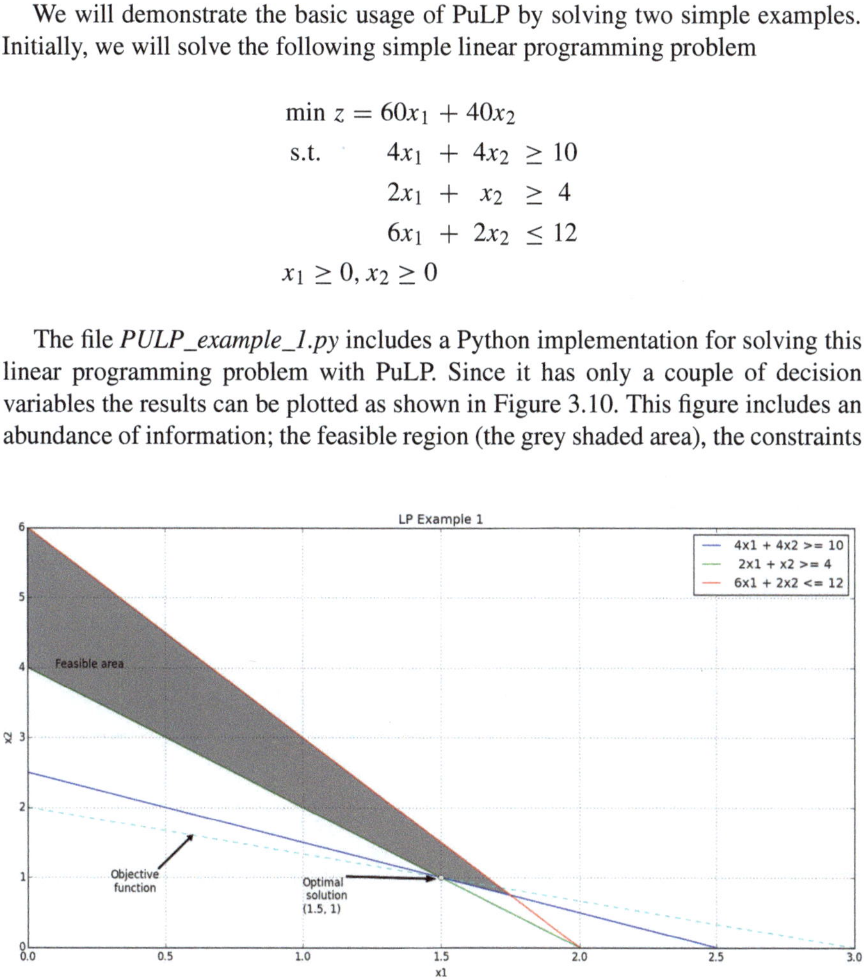

Fig. 3.10 Graphical solution of the first PuLP example

line segments, the model constraints, the optimal solution, and the line vertical to the objective function vector through the corner with the optimal objective function value found.

Initially, an object of a model is created (lines 11–12). Then, we define the decision variables (lines 15–16) as non-negative continuous variables. Next, we define the objective function (line 19) and the constraints (lines 22–24) of the problem. Finally, we solve the linear programming problem (line 27), then print (lines 30–36) and plot (lines 39–65) the results.

```
1.   # Filename: PULP_example_1.py
2.   # Description: An example of solving a linear
3.   # programming problem with PuLP
4.   # Authors: Papathanasiou, J. & Ploskas, N.
5.
6.   from pulp import *
7.   import matplotlib.pyplot as plt
8.   import numpy as np
9.
10.  # Create an object of a model
11.  prob = LpProblem("LP example with 2 decision "
12.      "variables", LpMinimize)
13.
14.  # Define the decision variables
15.  x1 = LpVariable("x1", 0)
16.  x2 = LpVariable("x2", 0)
17.
18.  # Define the objective function
19.  prob += 60*x1 + 40*x2
20.
21.  # Define the constraints
22.  prob += 4*x1 + 4*x2 >= 10.0, "1st constraint"
23.  prob += 2*x1 + x2 >= 4.0, "2nd constraint"
24.  prob += 6*x1 + 2*x2 <= 12.0, "3rd constraint"
25.
26.  # Solve the linear programming problem
27.  prob.solve()
28.
29.  # Print the results
30.  print ("Status: ", LpStatus[prob.status])
31.
32.  for v in prob.variables():
33.      print (v.name, "=", v.varValue)
34.
35.  print ("The optimal value of the objective function "
36.      "is = ", value(prob.objective))
37.
38.  # Plot the optimal solution
39.  x = np.arange(0, 5)
40.
41.  plt.plot(x, 2.5 - x, label = '4x1 + 4x2 >= 10')
42.  plt.plot(x, 4 - 2 * x, label= ' 2x1 + x2 >= 4')
```

```
plt.plot(x, 6 - 3 * x, label = '6x1 + 2x2 <= 12')
plt.plot(x, 2 - 2/3*x, '--')
plt.annotate('Objective\n function', xy = (0.6, 1.6),
    xytext = (0.3, 0.8), arrowprops =
    dict(facecolor = 'black', width = 1.5, headwidth = 7))
plt.annotate('Optimal\n solution\n(1.5, 1)',
    xy = (1.48, 0.98), xytext = (1, 0.5), arrowprops =
    dict(facecolor = 'black', width = 1.5, headwidth = 7))
plt.plot(1.5, 1, 'wo')
plt.text(0.1, 4, 'Feasible area', size = '12')

# Define the boundaries of the feasible area in the plot
a = [0, 1.5, 1.75, 0, 0]
b = [4, 1, 0.75, 6, 4]
plt.fill(a, b, 'grey')

plt.xlabel("x1")
plt.ylabel("x2")
plt.title('LP Example 1')
plt.axis([0, 3, 0, 6])
plt.grid(True)
plt.legend()
plt.show()
```

The output of the code in this case (the solution) is as follows:

```
Status: Optimal
x1 = 1.5
x2 = 1.0
The optimal value of the objective function is = 130.0
```

PuLP also solves mixed integer and binary linear programming problems. We will solve the following binary linear programming problem

$$
\begin{aligned}
\min z = {} & x_1 + x_2 + x_3 + x_4 + x_5 + x_6 \\
\text{s.t.} \quad & x_1 + x_3 \geq 2 \\
& x_1 + x_2 + x_5 \geq 2 \\
& x_3 + x_4 \geq 1 \\
& x_1 + x_4 + x_5 \geq 1 \\
& x_4 + x_5 + x_6 \geq 1 \\
x_i \in {} & \{0, 1\}, i = 1, 2, \cdots, 6
\end{aligned}
$$

The file *PULP_example_2.py* includes a Python implementation for solving this linear programming problem with PuLP. The only difference from the previous example is that now the programmer needs to explicitly declare that he/she will be using binary variables in the variable definition function (lines 15–20). Initially, an object of a model is created (lines 11–12). Then, we define the decision variables (lines 15–20) as binary variables. Next, we define the objective function (line 23) and the constraints (lines 26–30) of the problem. Finally, we solve the linear programming problem (line 33), and print the results (lines 36–42).

```
# Filename: PULP_example_1.py
# Description: An example of solving a binary
# linear programming problem with Pulp
# Authors: Papathanasiou, J. & Ploskas, N.

from pulp import *
import matplotlib.pyplot as plt
import numpy as np

# Create an object of a model
prob = LpProblem("Binary LP example with 6 decision "
    "variables", LpMinimize)

# Define the decision variables
x1 = LpVariable("x1", 0, 1, LpBinary)
x2 = LpVariable("x2", 0, 1, LpBinary)
x3 = LpVariable("x3", 0, 1, LpBinary)
x4 = LpVariable("x4", 0, 1, LpBinary)
x5 = LpVariable("x5", 0, 1, LpBinary)
x6 = LpVariable("x6", 0, 1, LpBinary)

# Define the objective function
prob += x1 + x2 + x3 + x4 + x5 + x6

# Define the constraints
prob += x1 + x3 >= 2
prob += x1 + x2 + x5 >= 2
prob += x3 + x4 >= 1
prob += x1 + x4 + x5 >= 1
prob += x4 + x5 + x6 >= 1

# Solve the linear programming problem
prob.solve()

# Print the results
print ("Status: ", LpStatus[prob.status])

for v in prob.variables():
    print (v.name, "=", v.varValue)

print ("The optimal value of the objective function "
    "is = ", value(prob.objective))
```

The output of the code in this case (the solution) is as follows:

```
Status: Optimal
x1 = 1.0
x2 = 0.0
x3 = 1.0
x4 = 0.0
x5 = 1.0
x6 = 0.0
The optimal value of the objective function is = 3.0
```

3.4.1 Numerical Example

Suppose that in the site selection example the decision maker needs to find a set of three sites and not just the best ranked solution. The easiest answer to this problem would be to select the three best ranked actions, but that may sometimes be misleading. If the investment criterion was not to be maximized but the other way around (this is not a public company now, but rather a private one), then there could be budget constraints. In this example, the three best ranked actions (6, 2, and 4) have a total investment of 20 million € (Table 3.3). If the overall available budget was a mere 18 million €, then those sites cannot be selected; another set must be selected and this is the case where PROMETHEE V comes in handy. With only this modification to the model, the results are shown in Table 3.13 and the ranking is shown in Figure 3.11.

The exact model is

$$
\begin{array}{llcccccccccccl}
\min z = & -0.42x_1 & + & 0.12x_2 & - & 0.06x_3 & + & 0.04x_4 & - & 0.1x_5 & + & 0.42x_6 & & \\
\text{s.t.} & x_1 & + & x_2 & + & x_3 & + & x_4 & + & x_5 & + & x_6 & = & 3 \\
 & x_1 & & & & & + & x_4 & & & & & \leq & 1 \\
 & & & x_2 & & & + & x_4 & & & & & \leq & 1 \\
 & 8x_1 & + & 5x_2 & + & 7x_3 & + & 9x_4 & + & 11x_5 & + & 6x_6 & \leq & 18 \\
\multicolumn{14}{l}{x_i \in \{0, 1\}, i = 1, 2, \cdots, 6}
\end{array}
$$

The first of the above constraints implies that only three out of six sites will be selected; the second that either site 1 or site 4 (or none) will be selected for whatever reason; the third has the same rationale, meaning either site 2 or site 4 (or none) will be selected; and the final constraint is about the overall available budget being 18 million €.

Table 3.13 Global preference net flows (minimization of investment criterion)

	Net flow (Φ)
Site 1	−0.42
Site 2	0.12
Site 3	−0.06
Site 4	0.04
Site 5	−0.1
Site 6	0.42

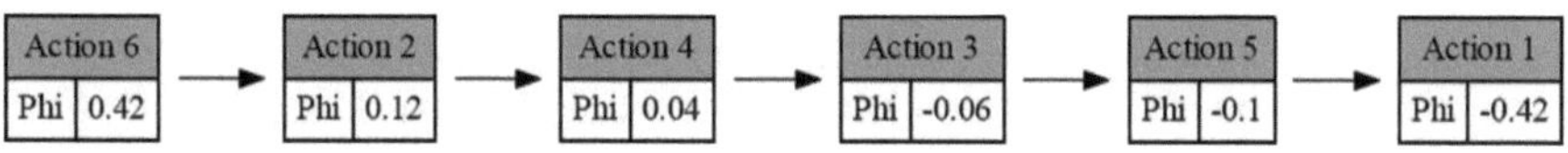

Fig. 3.11 Final ranking of the modified model

Solving this problem with an integer linear programming solver, like CBC (open source mixed integer programming solver)[5] or CPLEX, we derive the solution $x_1 = 0$, $x_2 = 1$, $x_3 = 1$, $x_4 = 0$, $x_5 = 0$, and $x_6 = 1$. Therefore, sites 2, 3, and 6 are selected with a total investment of 18 million €. If this amount is decreased further and the rest of the constraints remain the same, then the problem becomes infeasible.

3.4.2 Python Implementation

The file *PROMETHEE_V.py* includes a Python implementation of the PROMETHEE V method for the example presented in the previous subsection. Comments embedded in the code listing describe each part of the code. Initially, we call PROMETHEE II method to calculate the net flows (line 9). Next, an object of a model is created (lines 12). Then, we define the decision variables (lines 15–20) as binary variables. Next, we define the objective function (line 23–24) and the constraints (lines 27–31) of the problem. Finally, we solve the linear programming problem (line 34), and print the results (lines 37–43).

```
1.   # Filename: PROMETHEE_V.py
2.   # Description: PROMETHEE V method
3.   # Authors: Papathanasiou, J. & Ploskas, N.
4.
5.   from pulp import *
6.   from PROMETHEE_II import promethee, main
7.
8.   # Call PROMETHEE II to calculate the flows
9.   flows = main('n', 'n')
10.
11.  # Create an object of a model
12.  prob = LpProblem("Promethee V", LpMaximize)
13.
14.  # Define the decision variables
15.  x1 = LpVariable("x1", 0, 1, LpBinary)
16.  x2 = LpVariable("x2", 0, 1, LpBinary)
17.  x3 = LpVariable("x3", 0, 1, LpBinary)
18.  x4 = LpVariable("x4", 0, 1, LpBinary)
19.  x5 = LpVariable("x5", 0, 1, LpBinary)
20.  x6 = LpVariable("x6", 0, 1, LpBinary)
21.
22.  # Define the objective function
23.  prob += flows[0] * x1 + flows[1] * x2 + flows[2] \
24.      + flows[3] * x4 + flows[4] * x5 + flows[5] * x6
25.
26.  # Define the constraints
27.  prob += x1 + x2 + x3 + x4 + x5 + x6 == 3
```

[5] https://projects.coin-or.org/Cbc

```
prob += x1 + x4 <= 1
prob += x2 + x4 <= 1
prob += 8 * x1 + 5 * x2 + 7 * x3 + 9 * x4 + \
    11 * x5 + 6 * x6 <= 18

# Solve the linear programming problem
prob.solve()

# Print the results
print("Status:", LpStatus[prob.status])

for v in prob.variables():
    print(v.name, "=", v.varValue)

print("The optimal value of the objective function is = ",
    value(prob.objective))
```

The output of the code in this case (the solution) is as follows:

```
Status: Optimal
x1 = 0.0
x2 = 1.0
x3 = 1.0
x4 = 0.0
x5 = 0.0
x6 = 1.0
The optimal value of the objective function is = 0.48
```

References

1. Abu-Taleb, M. F., & Mareschal, B. (1995). Water resources planning in the Middle East: Application of the PROMETHEE V multicriteria method. *European Journal of Operational Research, 81*(3), 500–511.
2. Albuquerque, P. H. M. (2015). PROMETHEE IV as a decision analyst's tool for site selection in civil engineering. In P. Guarnieri (Ed.), *Decision models in engineering and management* (pp. 257–267). Cham: Springer International Publishing.
3. Antoniou, F., Aretoulis, G. N., Konstantinidis, D., & Papathanasiou, J. (2014). Choosing the most appropriate contract type for compensating major highway project contractors. *Journal of Computational Optimization in Economics and Finance, 6*(2), 77–95.
4. Aretoulis, G. N., Triantafyllidis, C. H., Papathanasiou, J., & Anagnostopoulos, I. K. (2015). Selection of the most competent project designer based on multi-criteria and cluster analysis. *International Journal of Data Analysis Techniques and Strategies, 7*(2), 172–186.
5. Bilsel, R. U., Buyukozkan, G., & Ruan, D. (2006). A fuzzy preference-ranking model for a quality evaluation of hospital web sites. *International Journal of Intelligent Systems, 21*(11), 1181–1197.
6. Brans, J. P., Macharis, C., & Mareschal, B. (1988). The GDSS PROMETHEE procedure (a PROMETHEE-GAIA based procedure for group decision support). *Journal of Decision Systems, 7*, 283–307.
7. Brans, J. P., & Mareschal, B. (1992). PROMETHEE V: MCDM problems with segmentation constraints. *Information Systems and Operational Research, 30*(2), 85–96.

8. Brans, J. P., & Mareschal, B. (1994). The PROMCALC & GAIA decision support system for multicriteria decision aid. *Decision Support Systems, 12*(4–5), 297–310.
9. Brans, J. P., & Mareschal, B. (1995). The PROMETHEE VI procedure: How to differentiate hard from soft multicriteria problems. *Journal of Decision Systems, 4*(3), 213–223.
10. Brans, J. P., & Mareschal, B. (2005). PROMETHEE methods. In J. Figueira, S. Greco, & M. Ehrgott (Eds.), *Multiple criteria decision analysis: State of the art surveys* (pp. 163–186). New York: Springer Science + Business Media, Inc.
11. Brans, J. P., Mareschal, B., & Vincke, P. (1984). PROMETHEE: A new family of outranking methods in multicriteria analysis. In J. P. Brans (Ed.), *Operational research '84* (pp. 477–490). Amsterdam: North Holland.
12. Brans, J. P., & Vincke, P. (1985). Note - A preference ranking organisation method (the PROMETHEE method for multiple criteria decision-making). *Management Science, 31*(6), 647–656.
13. Brans, J. P., Vincke, P., & Mareschal, B. (1986). How to select and how to rank projects: The PROMETHEE method. *European Journal of Operational Research, 24*(2), 228–238.
14. Briggs, T., Kunsch, P. L., & Mareschal, B. (1990). Nuclear waste management: An application of the multicriteria PROMETHEE methods. *European Journal of Operational Research, 44*(1), 1–10.
15. Cavalcante, C. A. V., & De Almeida, A. T. (2007). A multi-criteria decision-aiding model using PROMETHEE III for preventive maintenance planning under uncertain conditions. *Journal of Quality in Maintenance Engineering, 13*(4), 385–397.
16. Cavallaro, F. (2009). Multi-criteria decision aid to assess concentrated solar thermal technologies. *Renewable Energy, 34*(7), 1678–1685.
17. Chen, T. Y. (2014). A PROMETHEE-based outranking method for multiple criteria decision analysis with interval type-2 fuzzy sets. *Soft Computing, 18*(5), 923–940.
18. Chen, T. Y. (2015). IVIF-PROMETHEE outranking methods for multiple criteria decision analysis based on interval-valued intuitionistic fuzzy sets. *Fuzzy Optimization and Decision Making, 14*(2), 173–198.
19. Chen, Y. H., Wang, T. C., & Wu, C. Y. (2011). Strategic decisions using the fuzzy PROMETHEE for IS outsourcing. *Expert Systems with Applications, 38*(10), 13216–13222.
20. Chou, W. C., Lin, W. T., & Lin, C. Y. (2007). Application of fuzzy theory and PROMETHEE technique to evaluate suitable ecotechnology method: A case study in Shihmen Reservoir Watershed, Taiwan. *Ecological Engineering, 31*(4), 269–280.
21. de Almeida, A. T., & Vetschera, R. (2012). A note on scale transformations in the PROMETHEE V method. *European Journal of Operational Research, 219*(1), 198–200.
22. de Almeida, J. A., de Almeida, A. T., & Costa, A. P. (2014). Portfolio selection of information systems projects using PROMETHEE V with C-optimal concept. *Pesquisa Operacional, 34*(2), 275–299.
23. Dede, G., Kamalakis, T., & Sphicopoulos, T. (2015). Convergence properties and practical estimation of the probability of rank reversal in pairwise comparisons for multi-criteria decision making problems. *European Journal of Operational Research, 241*(2), 458–468.
24. Fernández-Castro, A. S. (2002). Capital budgeting in health organizations: Application of the multicriteria method PROMETHEE V. *Fuzzy Economic Review, 7*(2), 93–107.
25. Fernández-Castro, A. S., & Jiménez, M. (2005). PROMETHEE: An extension through fuzzy mathematical programming. *Journal of the Operational Research Society, 56*(1), 119–122.
26. Fontana, M. E., & Morais, D. C. (2013). Using PROMETHEE V to select alternatives so as to rehabilitate water supply network with detected leaks. *Water Resources Management, 27*(11), 4021–4037.
27. Geldermann, J., Spengler, T., & Rentz, O. (2000). Fuzzy outranking for environmental assessment. Case study: Iron and steel making industry. *Fuzzy Sets and Systems, 115*(1), 45–65.
28. Goumas, M., & Lygerou, V. (2000). An extension of the PROMETHEE method for decision making in fuzzy environment: Ranking of alternative energy exploitation projects. *European Journal of Operational Research, 123*(3), 606–613.

29. Gupta, R., Sachdeva, A., & Bhardwaj, A. (2012). Selection of logistic service provider using fuzzy PROMETHEE for a cement industry. *Journal of Manufacturing Technology Management, 23*(7), 899–921.
30. Ishizaka, A., & Nemery, P. (2013). *Multi-criteria decision analysis: Methods and software*. Hoboken: John Wiley & Sons.
31. Kaya, A. O., Kaya, T., & Kahraman, C. (2013). A fuzzy approach to urban ecotourism site selection based on an integrated PROMETHEE III methodology. *Journal of Multiple-Valued Logic & Soft Computing, 21*(1–2), 89–111.
32. Koutroumanidis, T., Papathanasiou, J., & Manos, B. (2002). A multicriteria analysis of productivity of agricultural regions of Greece. *Operational Research, 2*(3), 339–346.
33. Le Téno, J. F., & Mareschal, B. (1998). An interval version of PROMETHEE for the comparison of building products' design with ill-defined data on environmental quality. *European Journal of Operational Research, 109*(2), 522–529.
34. Li, W. X., & Li, B. Y. (2010). An extension of the PROMETHEE II method based on generalized fuzzy numbers. *Expert Systems with Applications, 37*(7), 5314–5319.
35. López, H. M. L., & Almeida, A. T. D. (2014). Project portfolio selection in an electric utility company using PROMETHEE V. *Production, 24*(3), 559–571.
36. Mareschal, B. (2017). *PROMETHEE - GAIA Statistics*. http://www.promethee-gaia.net/assets/promethee-stats.pdf (Current as of 30 April 2017).
37. Mareschal, B., De Smet, Y., & Nemery, P. (2008). Rank reversal in the PROMETHEE II method: Some new results. In *Proceedings of the 2008 IEEE International Conference on Industrial Engineering and Engineering Management, Singapore* (pp. 959–963).
38. Mavrotas, G., Diakoulaki, D., & Caloghirou, Y. (2006). Project prioritization under policy restrictions. A combination of MCDA with 0–1 programming. *European Journal of Operational Research, 171*(1), 296–308.
39. Motlagh, S. M. H., Behzadian, M., Ignatius, J., Goh, M., Sepehri, M. M., & Hua, T. K. (2015). Fuzzy PROMETHEE GDSS for technical requirements ranking in HOQ. *The International Journal of Advanced Manufacturing Technology, 76*(9–12), 1993–2002.
40. Opricovic, S., & Tzeng, G. H. (2007). Extended VIKOR method in comparison with outranking methods. *European Journal of Operational Research, 178*(2), 514–529.
41. PuLP (2016). https://pypi.python.org/pypi/PuLP (Current as of 30 April 2017).
42. Senvar, O., Tuzkaya, G., & Kahraman, C. (2014). Multi criteria supplier selection using fuzzy PROMETHEE method. *Studies in Fuzziness and Soft Computing, 313*, 21–34.
43. Yatsalo, B., Didenko, V., Gritsyuk, S., & Sullivan, T. (2015). Decerns: A framework for multi-criteria decision analysis. *International Journal of Computational Intelligence Systems, 8*(3), 467–489.
44. Zhang, K., Kluck, C., & Achari, G. (2009). A comparative approach for ranking contaminated sites based on the risk assessment paradigm using fuzzy PROMETHEE. *Environmental Management, 44*(5), 952–967.

Chapter 4
SIR

4.1 Introduction

The superiority and inferiority ranking method (SIR) is a relatively new MCDA methodology. SIR was initially proposed by Xu [14]. This chapter will be based on his research since SIR is a new entry in the MCDA anthology and there exist only a few references to this methodology. It is presented in this book mainly because it is closely related to TOPSIS and PROMETHEE that were presented in previous chapters; indeed according to Xu [14], SIR is an extension of PROMETHEE. It has been applied to domains like contractor selection [7], procurement decisions [8], site selection [13], job evaluation [12], concrete supply with stochastic input data [5], and selection of solar energy [4]. Furthermore, the method has been extended and refined to group decision making with fuzzy numbers [3, 6]. Alternative aggregation approaches using grey theory have been proposed by Tam and Tong [13].

4.2 Methodology

As in the previous chapters, initially it is necessary to define the problem's exact parameters; these are m alternatives $A_1, A_2, \cdots, A_m$ and n cardinal criteria $g_1, g_2, \cdots, g_n$. Let $g_j(A_i)$ be the performance of the ith alternative A_i with respect to the jth criterion g_j, where $1 \leq i \leq m$ and $1 \leq j \leq n$. Without loss of generality, let us also assume that all criteria are to be maximized. This leads to the formulation of the decision matrix D as follows:

Electronic Supplementary Material The online version of this chapter (https://doi.org/10.1007/978-3-319-91648-4_4) contains supplementary material, which is available to authorized users.

J. Papathanasiou, N. Ploskas, *Multiple Criteria Decision Aid*, Springer Optimization and Its Applications 136, https://doi.org/10.1007/978-3-319-91648-4_4

$$D = \begin{pmatrix} g_1(A_1) & \cdots & g_j(A_1) & \cdots & g_n(A_1) \\ \cdots & \cdots & \cdots & \cdots & \cdots \\ g_1(A_i) & \cdots & g_j(A_i) & \cdots & g_n(A_i) \\ \cdots & \cdots & \cdots & \cdots & \cdots \\ g_1(A_m) & \cdots & g_j(A_m) & \cdots & g_n(A_m) \end{pmatrix} \tag{4.1}$$

For the weights w_j $(j = 1, 2, \cdots, n)$, the following equation is applied (as in previous chapters).

$$\sum_{j=1}^{n} w_j = 1(w_j \geq 0) \tag{4.2}$$

Rebai [9, 10] defined the concepts of superiority, inferiority, noninferiority, and indifference scores (SINE scores); Xu [14] generalized these by using ideas from the PROMETHEE methods. The intensity of preference $P(A, A')$ of alternative A over A' on criterion g is denoted as follows:

$$P(A, A') = f(g(A) - g(A')) = f(d) \tag{4.3}$$

where $f(d)$ is a non-decreasing function from $\mathbb{R}$ to [0, 1] such that $f(d) = 0$ for $d \leq 0$. There are six generalized criteria as in [1] (see Figure 3.1). The same rules that apply in PROMETHEE for the definitions of the q and p parameters (indifference and absolute preference thresholds, respectively) apply here as well.

For each of the alternatives, the superiority index $S_j(A_i)$ and inferiority index $I_j(A_i)$ are defined by the following formulas:

$$S_j(A_i) = \sum_{k=1}^{m} P_j(A_i, A_k) = \sum_{k=1}^{m} f_j(g_j(A_i) - g_j(A_k)) \tag{4.4}$$

$$I_j(A_i) = \sum_{k=1}^{m} P_j(A_k, A_i) = \sum_{k=1}^{m} f_j(g_j(A_k) - g_j(A_i)) \tag{4.5}$$

Therefore, the superiority matrix (S-matrix) is as

$$S = \begin{pmatrix} S_1(A_1) & \cdots & S_j(A_1) & \cdots & S_n(A_1) \\ \cdots & \cdots & \cdots & \cdots & \cdots \\ S_1(A_i) & \cdots & S_j(A_i) & \cdots & S_n(A_i) \\ \cdots & \cdots & \cdots & \cdots & \cdots \\ S_1(A_m) & \cdots & S_j(A_m) & \cdots & S_n(A_m) \end{pmatrix} \text{ or } S = (S_j(A_i))_{m \times n} \tag{4.6}$$

and the inferiority matrix (I-matrix) as

$$I = \begin{pmatrix} I_1(A_1) & \cdots & I_j(A_1) & \cdots & I_n(A_1) \\ \cdots & \cdots & \cdots & \cdots & \cdots \\ I_1(A_i) & \cdots & I_j(A_i) & \cdots & I_n(A_i) \\ \cdots & \cdots & \cdots & \cdots & \cdots \\ I_1(A_m) & \cdots & I_j(A_m) & \cdots & I_n(A_m) \end{pmatrix} \text{ or } I = (I_j(A_i))_{m \times n} \tag{4.7}$$

These matrices contain different kinds of information as the comparison is not the same; the S-matrix is about the intensity of superiority of each alternative on each criterion, while the I-matrix is about the intensity of inferiority of each alternative on each criterion.

Once the S-matrix and I-matrix are calculated, they are aggregated into two types of global preference indices, the superiority flow (S-flow) $\phi^{>}(\cdot)$ and the inferiority flow (I-flow) $\phi^{<}(\cdot)$. The former represents the global intensity of superiority for each alternative, while the latter the global intensity of inferiority. For each alternative, if V is defined as the aggregation function, the superiority flow is defined as

$$\phi^{>}(A_i) = V\left[S_1(A_i), \cdots, S_j(A_i), \cdots, S_n(A_i)\right] \tag{4.8}$$

As mentioned, the superiority flow $\phi^{>}(A_i)$ is a measurement of how A_i globally outranks all the other alternatives.

The inferiority flow is as

$$\phi^{<}(A_i) = V\left[I_1(A_i), \cdots, I_j(A_i), \cdots, I_n(A_i)\right] \tag{4.9}$$

The inferiority flow $\phi^{<}(A_i)$ is a measurement of how A_i globally is outranked by all the other alternatives. Clearly, the higher S-flow and the lower I-flow, the better is A_i.

Xu [14] proposes a couple of methodologies for the aggregation procedure; namely, SAW (Simple Additive Weighting) and TOPSIS. He considers however the list not to be exhaustive and encourages the decision maker to test other procedures as well (like in [13]).

Therefore, if the decision maker opts for SAW (SIR·SAW), the S-flow and I-flow are as follows:

$$\phi^{>}(A_i) = \sum_{j=1}^{n} w_j S_j(A_i) \tag{4.10}$$

and

$$\phi^{<}(A_i) = \sum_{j=1}^{n} w_j I_j(A_i) \tag{4.11}$$

This is where PROMETHEE comes in; Xu [14] provides proof that the S-flow and I-flow are the same as the leaving and entering flows, respectively, of PROMETHEE.

If the decision maker chooses the classical version of TOPSIS presented in Chapter 1 (SIR·TOPSIS, other TOPSIS versions can be used as well), then the ideal solution A_S^+ and the anti-ideal solution A_S^- for the superiority matrix $S = (S_j(A_i))_{m \times n}$ are

$$A_S^+ = \left(\max_i S_1(A_i), \max_i S_2(A_i), \cdots, \max_i S_n(A_i) \right) = (S_1^+, S_2^+ \cdots, S_n^+) \tag{4.12}$$

and

$$A_S^- = \left(\min_i S_1(A_i), \min_i S_2(A_i), \cdots, \min_i S_n(A_i) \right) = (S_1^-, S_2^-, \cdots, S_n^-) \tag{4.13}$$

The superiority flow is

$$\phi^{>}(A_i) = \frac{S^-(A_i)}{S^-(A_i) + S^+(A_i)} \tag{4.14}$$

where

$$S^+(A_i) = \left(\sum_{j=1}^{n} \left| w_j \left(S_j(A_i) - S_j^+ \right) \right|^{\lambda} \right)^{1/\lambda} \tag{4.15}$$

and

$$S^-(A_i) = \left(\sum_{j=1}^{n} \left| w_j \left(S_j(A_i) - S_j^- \right) \right|^{\lambda} \right)^{1/\lambda} \tag{4.16}$$

The Minkowski distance is used between two vectors $a = (a_1, a_2, \cdots, a_n)$ and $b = (b_1, b_2, \cdots, b_n)$.

$$d^{\lambda}(a, b) = \left(\sum_{j=1}^{n} \left| a_j - b_j \right|^{\lambda} \right)^{1/\lambda}, 1 \leq \lambda \leq \infty \tag{4.17}$$

It has to be noted that the following cases apply

- if $\lambda = 2$, Equation (4.17) corresponds to the Euclidean distance.
- if $\lambda = 1$, Equation (4.17) is as Equation (4.18) (block distance)

$$d^1(a, b) = \sum_{j=1}^{n} \left| a_j - b_j \right| \tag{4.18}$$

- if $\lambda = \infty$, Equation (4.17) is as Equation (4.19)

$$d^{\infty}(a, b) = \max_j \left| a_j - b_j \right| \tag{4.19}$$

In a similar fashion, the ideal and the anti-ideal solution for the inferiority matrix $I = \left(I_j(A_i)\right)_{m \times n}$ are defined as

$$A_I^+ = \left(\min_i I_1(A_i), \min_i I_2(A_i), \cdots, \min_i I_n(A_i) \right) = (I_1^+, I_2^+, I_n^+) \tag{4.20}$$

$$A_I^- = \left(\max_i I_1(A_i), \max_i I_2(A_i), \cdots, \max_i I_n(A_i) \right) = (I_1^-, I_2^-, I_n^-) \tag{4.21}$$

The inferiority flow is accordingly

$$\phi^{<}(A_i) = \frac{I^+(A_i)}{I^+(A_i) + I^-(A_i)} \tag{4.22}$$

where

$$I^+(A_i) = \left(\sum_{j=1}^{n} \left| w_j \left(I_j(A_i) - I_j^+ \right) \right|^{\lambda} \right)^{1/\lambda} \tag{4.23}$$

and

$$I^-(A_i) = \left(\sum_{j=1}^{n} \left| w_j \left(I_j(A_i) - I_j^- \right) \right|^{\lambda} \right)^{1/\lambda} \tag{4.24}$$

The same rules apply for λ in Equations (4.23) and (4.24) as in the cases of Equations (4.15) and (4.16). When the superiority $\phi^{>}(A_i)$ and inferiority $\phi^{<}(A_i)$ flows are obtained, then the ranking can be produced. The higher the superiority flow and the lower the inferiority flow at the same time, the better rank in the list for the alternative. The outcome of the method can be a partial or complete ranking. For the first case, the S-ranking $\mathbb{R}_{>} = \{P_{>}, I_{>}\}$ is a complete ranking of the alternatives according to the descending order of the superiority flow.

$$A_i P_{>} A_k \text{ iff } \phi^{>}(A_i) > \phi^{>}(A_k) \text{ and } A_i I_{>} A_k \text{ iff } \phi^{>}(A_i) = \phi^{>}(A_k) \tag{4.25}$$

A second complete ranking is produced according to the ascending order of $\phi^{<}(A_i)$ called I-ranking $\mathbb{R}_{<} = \{P_{<}, I_{<}\}$.

$$A_i P_{<} A_k \text{ iff } \phi^{<}(A_i) < \phi^{<}(A_k) \text{ and } A_i I_{<} A_k \text{ iff } \phi^{<}(A_i) = \phi^{<}(A_k) \tag{4.26}$$

Then the combination of $\mathbb{R}_{>} = \{P_{>}, I_{>}\}$ and $\mathbb{R}_{<} = \{P_{<}, I_{<}\}$ is a partial ranking $\mathbb{R} = \{P, I, R\} = \mathbb{R}_{>} \cap \mathbb{R}_{<}$ following the intersection principle [2, 11] that is defined given any two alternatives A and A' as

1. the preference relation P by

$$APA' \text{ iff } \left(AP_{>}A' \text{ and } AP_{<}A'\right) \text{ or } \left(AP_{>}A' \text{ and } AI_{<}A'\right) \text{ or } \left(AI_{>}A' \text{ and } AP_{<}A'\right) \tag{4.27}$$

2. the indifference relation I by

$$AIA' \text{ iff } AI_{>}A' \text{ and } AI_{<}A' \tag{4.28}$$

3. the incompatibility relation R by

$$ARA' \text{ iff } \left(AP_{>}A' \text{ and } AP_{<}A'\right) \text{ or } \left(A'P_{>}A \text{ and } AP_{<}A'\right) \tag{4.29}$$

The whole procedure of SIR partial ranking is depicted in Figure 4.1.

Figure 4.2 outlines the second case, i.e., the complete ranking achieved by SIR. In this case, the net flow (n-flow) is computed as (equivalent to the PROMETHEE net flow)

$$\phi_n(A_i) = \phi^{>}(A_i) - \phi^{<}(A_i) \tag{4.30}$$

and the relative flow as (r-flow, equivalent to the distance measurements of TOPSIS)

$$\phi_r(A_i) = \frac{\phi^{>}(A_i)}{\phi^{>}(A_i) + \phi^{<}(A_i)} \tag{4.31}$$

The final complete ranking $\mathbb{R}_n$ or $\mathbb{R}_r$ is obtained by $\phi_n(A_i)$ or $\phi_r(A_i)$, respectively.

SIR is closely related to PROMETHEE and TOPSIS; indeed Xu [14] proposes that (Figure 4.3)

1. PROMETHEE is equivalent to SIR·SAW
2. SAW is a special case of SIR·SAW

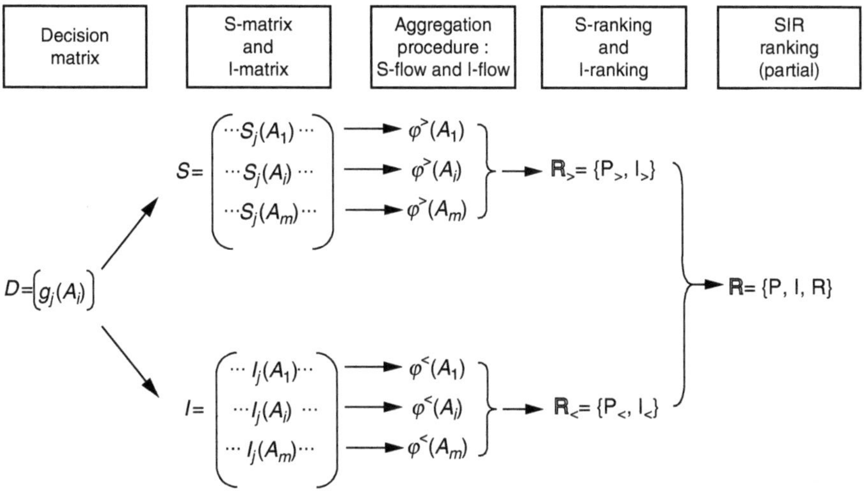

Fig. 4.1 The process of SIR method - partial ranking (adopted from [14])

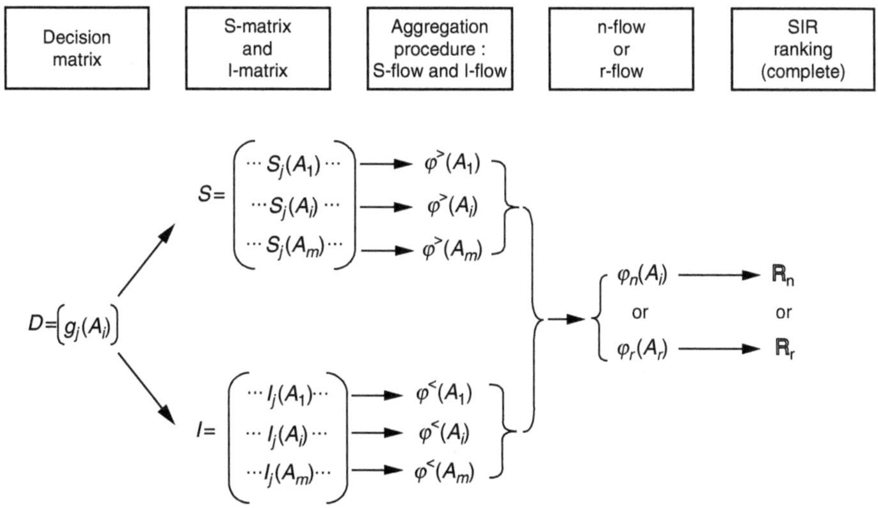

Fig. 4.2 The process of SIR method - complete ranking (adopted from [14])

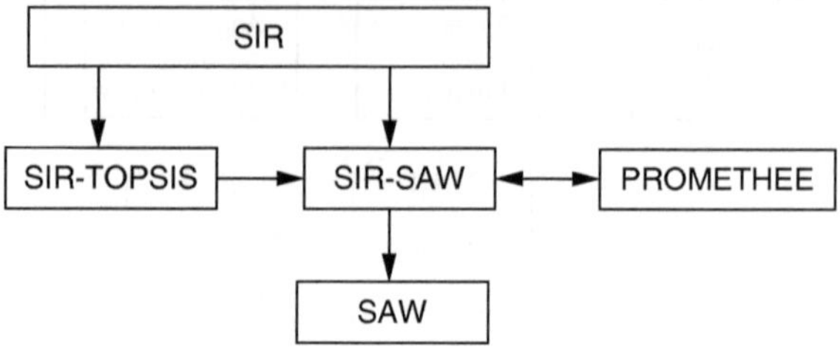

Fig. 4.3 The relationships of some MCDA methods (adopted from [14])

3. SAW is a special case of PROMETHEE
4. SIR·SAW is a special case of SIR·TOPSIS
5. PROMETHEE is a special case of SIR·TOPSIS

Concluding, the steps of the SIR method are

1. Build the decision matrix and set the values of the preference parameters and the weights
2. Calculate the S-matrix and I-matrix
3. Choose the proper aggregation procedure
   - SIR·SAW
   - SIR·TOPSIS
   - any other aggregation method
4. Choose the distance metric (λ value) and calculate the S-flow and I-flow
5. Rank the alternatives
   - either partially: calculate S-ranking and I-ranking
   - or completely: calculate n-flow or r-flow

4.2.1 *Numerical Example*

To allow the reader to compare SIR with other methods presented in previous chapters, the same example will be used. Table 1.2 is repeated here as Table 4.1. We assume a linear preference function for all criteria (where $q = 1$ and $p = 2$). Initially, we calculate the S-matrix and I-matrix using Equations (4.4) and (4.5), respectively. Tables 4.2 and 4.3 show the S-matrix and I-matrix, respectively. If the decision maker decides to use SAW as the aggregation procedure, then the flows are shown in Table 4.4. They can be obtained using Equations (4.10) (for S-flow, $\phi^{>}(A_i)$) and (4.11) (for I-flow, $\phi^{<}(A_i)$). Then, the n-flow ($\phi_n(A_i)$) is calculated using Equation (4.30) and the r-flow ($\phi_r(A_i)$) using Equation (4.31).

Table 4.1 Initial data

	Investment costs (million €)	Employment needs (hundred employees)	Social impact (1–7)	Environmental impact (1–7)
Weight	0.4	0.4	0.1	0.2
Site 1	8	7	2	1
Site 2	5	3	7	5
Site 3	7	5	6	4
Site 4	9	9	7	3
Site 5	11	10	3	7
Site 6	6	9	5	4

Table 4.2 The superiority matrix (S-matrix)

	Investment costs	Employment needs	Social impact	Environmental impact
Site 1	2	2	0	0
Site 2	0	0	3	2
Site 3	1	1	2	1
Site 4	3	3	3	1
Site 5	5	3	0	5
Site 6	0	3	2	1

Table 4.3 The inferiority matrix (I-matrix)

	Investment costs	Employment needs	Social impact	Environmental impact
Site 1	1	3	4	5
Site 2	4	5	0	1
Site 3	2	4	0	1
Site 4	1	0	0	2
Site 5	0	0	4	0
Site 6	3	0	2	1

Table 4.4 The flows of SIR·SAW

S-flow	I-flow	n-flow	r-flow
$\phi^{>}(A_1) = 1.400$	$\phi^{<}(A_1) = 2.700$	$\phi_n(A_1) = -1.300$	$\phi_r(A_1) = 0.341$
$\phi^{>}(A_2) = 0.700$	$\phi^{<}(A_2) = 3.300$	$\phi_n(A_2) = -2.600$	$\phi_r(A_2) = 0.175$
$\phi^{>}(A_3) = 1.100$	$\phi^{<}(A_3) = 2.200$	$\phi_n(A_3) = -1.100$	$\phi_r(A_3) = 0.333$
$\phi^{>}(A_4) = 2.600$	$\phi^{<}(A_4) = 0.800$	$\phi_n(A_4) = 1.800$	$\phi_r(A_4) = 0.765$
$\phi^{>}(A_5) = 3.900$	$\phi^{<}(A_5) = 0.400$	$\phi_n(A_5) = 3.500$	$\phi_r(A_5) = 0.907$
$\phi^{>}(A_6) = 1.300$	$\phi^{<}(A_6) = 1.600$	$\phi_n(A_6) = -0.300$	$\phi_r(A_6) = 0.448$

The two complete rankings are:

- $S - ranking$ - $\mathbb{R}_{>}$: $A_5 \rightarrow A_4 \rightarrow A_1 \rightarrow A_6 \rightarrow A_3 \rightarrow A_2$
- $I - ranking$ - $\mathbb{R}_{<}$: $A_5 \rightarrow A_4 \rightarrow A_6 \rightarrow A_3 \rightarrow A_1 \rightarrow A_2$

The resulting partial ranking is shown in Figure 4.4 (this is also the partial ranking of PROMETHEE I).

The complete ranking by n-flow is (this is also the complete ranking of PROMETHEE II):

$$\mathbb{R}_n: A_5 \rightarrow A_4 \rightarrow A_6 \rightarrow A_3 \rightarrow A_1 \rightarrow A_2$$

The complete ranking by r-flow is:

$$\mathbb{R}_r: A_5 \rightarrow A_4 \rightarrow A_6 \rightarrow A_1 \rightarrow A_3 \rightarrow A_2$$

Similarly, if the decision maker decides to use TOPSIS as the aggregation procedure, then the flows are shown in Table 4.5. They can be obtained using Equations (4.14) (for S-flow, $\phi^{>}(S_i)$) and (4.22) (for I-flow, $\phi^{<}(S_i)$). Then, the n-flow ($\phi_n(S_i)$) is calculated using Equation (4.30) and the r-flow ($\phi_r(S_i)$) using Equation (4.31). We also use the Euclidean distance, i.e., $\lambda = 2$.

The two complete rankings are:

- $S - ranking$ - $\mathbb{R}_{>}$: $A_5 \rightarrow A_4 \rightarrow A_1 \rightarrow A_6 \rightarrow A_3 \rightarrow A_2$
- $I - ranking$ - $\mathbb{R}_{<}$: $A_5 \rightarrow A_4 \rightarrow A_6 \rightarrow A_1 \rightarrow A_3 \rightarrow A_2$

The resulting partial ranking is shown in Figure 4.5.

The complete ranking by n-flow is:

Fig. 4.4 Resulting partial ranking using SIR·SAW

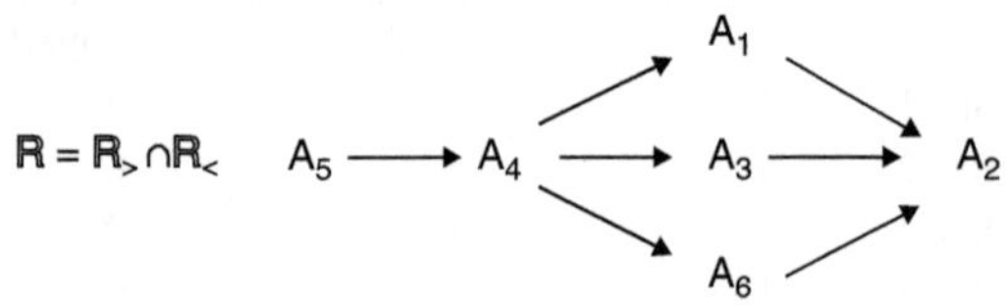

Table 4.5 The flows of SIR·TOPSIS

S-flow	I-flow	n-flow	r-flow
$\phi^{>}(A_1) = 0.382$	$\phi^{<}(A_1) = 0.521$	$\phi_n(A_1) = -0.139$	$\phi_r(A_1) = 0.423$
$\phi^{>}(A_2) = 0.180$	$\phi^{<}(A_2) = 0.711$	$\phi_n(A_2) = -0.531$	$\phi_r(A_2) = 0.202$
$\phi^{>}(A_3) = 0.233$	$\phi^{<}(A_3) = 0.541$	$\phi_n(A_3) = -0.308$	$\phi_r(A_3) = 0.301$
$\phi^{>}(A_4) = 0.577$	$\phi^{<}(A_4) = 0.216$	$\phi_n(A_4) = 0.361$	$\phi_r(A_4) = 0.727$
$\phi^{>}(A_5) = 0.889$	$\phi^{<}(A_5) = 0.142$	$\phi_n(A_5) = 0.747$	$\phi_r(A_5) = 0.862$
$\phi^{>}(A_6) = 0.304$	$\phi^{<}(A_6) = 0.412$	$\phi_n(A_6) = -0.108$	$\phi_r(A_6) = 0.425$

Fig. 4.5 Resulting partial ranking using SIR·TOPSIS

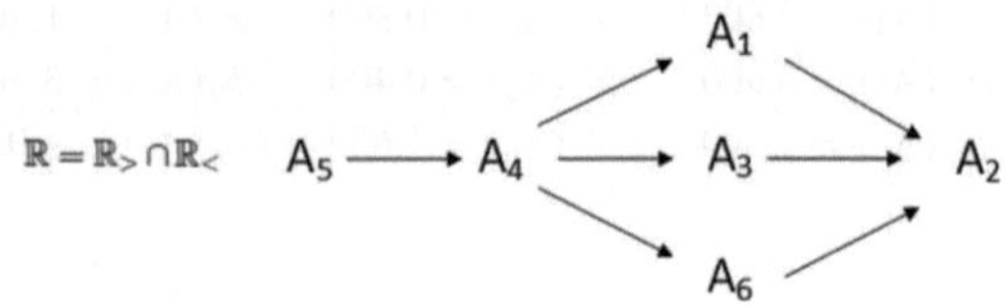

$$\mathbb{R}_n\colon A_5 \rightarrow A_4 \rightarrow A_6 \rightarrow A_1 \rightarrow A_3 \rightarrow A_2$$

The complete ranking by r-flow is:

$$\mathbb{R}_r\colon A_5 \rightarrow A_4 \rightarrow A_6 \rightarrow A_1 \rightarrow A_3 \rightarrow A_2$$

4.2.2 Python Implementation

The file *SIR.py* includes a Python implementation of SIR. The input variables are arrays x (action performances, lines 142–143), p (preference parameters of all criteria, line 146), *c1* (criteria optimization for calculating the S-matrix, line 150), *c2* (criteria optimization for calculating the I-matrix, line 154), d (preference functions, line 157), d (preference function array, line 157), and w (weights of criteria, line 160). The function *pref_func* calculates the preference degrees (lines 10–58), while the function *SImatrix* calculates S-matrix and I-matrix. The function *SIflowsSAW* calculates the S-flow and I-flow if SIR·SAW is used as the aggregation procedure. The function *SIRTOPSIS* calculates S^+ and I^+ and the function *SIflowsTOPSIS* calculates the S-flow and I-flow if SIR·TOPSIS is used as the aggregation procedure. The functions *Nflow* and *Rflow* calculate the n-flow and r-flow, respectively. The final results are displayed in lines 192–195.

```
1.  # Filename: SIR.py
2.  # Description: SIR method
3.  # Authors: Papathanasiou, J. & Ploskas, N.
4.
5.  from numpy import *
6.  import matplotlib.pyplot as plt
7.  from SIR_Final_Rank_Figure import graph, plot
8.
9.  # Calculate the preference degrees
10. def pref_func(a, b, c, d, e, m):
11.     """ a and b are action performances, c is q, d is p,
12.     e is the preference function ('u' for usual, 'us'
13.     for u-shape, 'vs' for v-shape, 'le' for level,
14.     'li' for linear, and 'g' for Gaussian), m is min/max
15.     """
16.     f = float(1.0)
17.     if m == 1:
18.         temp = a
19.         a = b
20.         b = temp
21.     if e == 'u': # Usual preference function
22.         if b - a > 0:
23.             f = 1
24.         else:
25.             f = 0
```

```
    elif e == 'us': # U-shape preference function
        if b - a > c:
            f = 1
        elif b - a <= c:
            f = 0
    elif e == 'vs': # V-shape preference function
        if b - a > d:
            f = 1
        elif b - a <= 0:
            f = 0
        else:
            f = (b - a) / d
    elif e == 'le': # Level preference function
        if b - a > d:
            f = 1
        elif b - a <= c:
            f = 0
        else:
            f = 0.5
    elif e == 'li': # Linear preference function
        if b - a > d:
            f = 1
        elif b - a <= c:
            f = 0
        else:
            f = ((b - a) - c) / (d - c)
    elif e == 'g': # Gaussian preference function
        if b - a > 0:
            f = 1 - math.exp(-(math.pow(b - a, 2)
                / (2 * d ** 2)))
        else:
            f = 0
    return f

# Calculate S and I matrices
def SImatrix(x, p, c, d):
    """ x is the action performances array, p is the
    array with the preference parameters of all criteria,
    c is the criteria min (0) or max (1) optimization
    array, and d is the preference function array for
    a specific criterion ('u' for usual, 'us' for u-shape,
    'vs' for v-shape, 'le' for level, 'li' for linear,
    and 'g' for Gaussian)
    """
    SI = zeros((size(x, 0), size(x, 1)))
    for i in range(size(x, 1)):
        for j in range(size(x, 0)):
            k = 0
            for h in range(size(x, 0)):
                k = k + pref_func(x[j, i], x[h, i],
                    p[0, i], p[1, i], d[i], c[i])
                SI[j, i] = k
    return SI

```

```
# Calculate S- and I-flow for SIR-SAW
def SIflowsSAW(w, SI):
    """ w is the weights array and SI is S or I matrix
    """
    k = zeros(size(SI, 0))
    for i in range(size(SI, 0)):
        for j in range(size(SI, 1)):
            k[i] = k[i] + w[j] * SI[i, j]
    return k

# Calculate SIplus and SIminus for SIR-TOPSIS
def SIRTOPSIS(w, SI, l):
    """ w is the weights array, SI is S or I matrix,
    and l is the distance metric
    """
    SIplus = zeros((size(SI, 0)))
    SIminus = zeros((size(SI, 0)))
    bb = []
    cc = []
    for i in range((size(w, 0))):
        bb.append(amax(SI[:, i:i + 1]))
        bbb = array(bb)
        cc.append(amin(SI[:, i:i + 1]))
        ccc = array(cc)
    for i in range((size(SI, 0))):
        for j in range((size(SI, 1))):
            SIplus[i] = SIplus[i] + math.pow(w[j]
                * abs(SI[i, j] - bbb[j]), l)
            SIminus[i] = SIminus[i] + math.pow(w[j]
                * abs(SI[i, j] - ccc[j]), l)
        SIplus[i] = math.pow(SIplus[i], 1 / l)
        SIminus[i] = math.pow(SIminus[i], 1 / l)
    return SIplus, SIminus

# Calculate the S- and I-flow for SIR-TOPSIS
def SIflowsTOPSIS(p, m):
    """ p is SIplus and m is SIminus
    """
    return m / (m + p)

# Calculate n-flow
def Nflow(s, i):
    """ s is S-flow and i is I-flow
    """
    return s - i

# Calculate r-flow
def Rflow(s, i):
    """ a is S-flow and b is I-flow
    """
    return s / (s + i)

# main function
def main(a, b, c):
```

```
    """ a, b, and c are flags; if a and b are set to
    'y' they do print the results, anything else does
    not print the results. If c equals 1, SIR-SAW is
    used as the aggregation procedure; otherwise,
    SIR-TOPSIS is used
    """

    # action performances array
    x = array([[8, 7, 2, 1], [5, 3, 7, 5], [7, 5, 6, 4],
        [9, 9, 7, 3], [11, 10, 3, 7], [6, 9, 5, 4]])

    # preference parameters of all criteria array
    p = array([[1, 1, 1, 1], [2, 2, 2, 2]])

    # criteria min (0) or max (1) optimization array for
    # calculating S matrix
    c1 = ([1, 1, 1, 1])

    # criteria min (0) or max (1) optimization array for
    # calculating I matrix (the opposite of c1)
    c2 = ([0, 0, 0, 0])

    # preference function array
    d = (['li', 'li', 'li', 'li'])

    # weights of criteria
    w = array([0.4, 0.3, 0.1, 0.2])

    # calculate S matrix
    S = SImatrix(x, p, c1, d)
    print("S = ", S)

    # calculate I matrix
    I = SImatrix(x, p, c2, d)
    print("I = ", I)

    if c == 1: # SIR-SAW
        # calculate S-flow
        Sflow = SIflowsSAW(w, S)

        # calculate I-flow
        Iflow = SIflowsSAW(w, I)
    else: # SIR-TOPSIS
        # calculate S-flow
        Splus, Sminus = SIRTOPSIS(w, S, 2)
        Sflow = SIflowsTOPSIS(Splus, Sminus)

        # calculate I-flow
        Iplus, Iminus = SIRTOPSIS(w, I, 2)
        Iflow = SIflowsTOPSIS(Iplus, Iminus)

    # calculate n-flow
    nflow = Nflow(Sflow, Iflow)

```

```
    # calculate r-flow
    rflow = Rflow(Sflow, Iflow)

    # print flows
    print("S-flow = ", Sflow)
    print("I-flow = ", Iflow)
    print("n-flow = ", nflow)
    print("r-flow = ", rflow)

    # plot results
    if a == 'y':
        graph(around(rflow, 3), "Flow")
    if b == 'y':
        plot(around(rflow, 3), "SIR")

if __name__ == '__main__':
    main('n', 'y', 1)
```

The file *SIR_Final_Rank_Figure.py* is an optional module that includes two methods to plot the results of SIR. This module produces two plots with the results (Figures 4.6 and 4.7 display the results when SIR·SAW is used as the aggregation method). In order to run these methods, one needs to install graphviz and matplotlib modules.

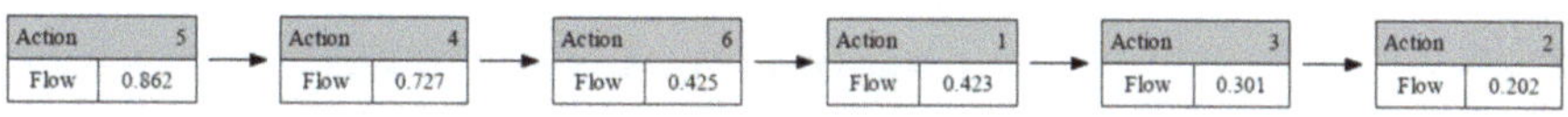

Fig. 4.6 Final rank figure plotted with graphviz module

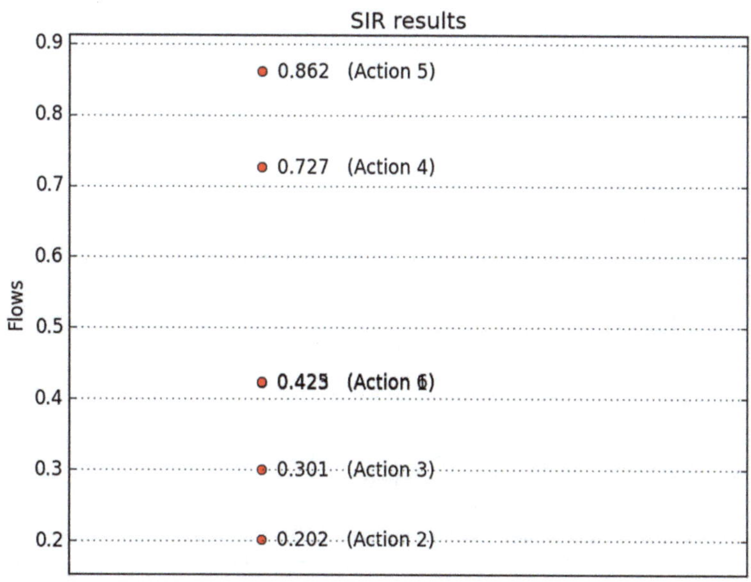

Fig. 4.7 Final rank plotted with matplotlib

```
# Filename: SIR_Final_Rank_Figure.py
# Description: Optional module to plot the
# results of SIR method
# Authors: Papathanasiou, J. & Ploskas, N.

import matplotlib.pyplot as plt
from graphviz import Digraph
from numpy import *

# Plot final rank figure
def graph(flows, b):
    """ flows is the matrix with the flows, and b
    is a string describing the flow
    """
    s = Digraph('Actions', node_attr = {'shape':
        'plaintext'})
    s.body.extend(['rankdir = LR'])
    x = sort(flows)
    y = argsort(flows)
    l = []
    for i in y:
        s.node('action' + str(i), '''<
        <TABLE BORDER="0" CELLBORDER="1"
            CELLSPACING="0" CELLPADDING="4">
          <TR>
            <TD COLSPAN="2" bgcolor="grey" >Action
                ''' + str(y[i] + 1) + '''</TD>
          </TR>
          <TR>
            <TD>'''+ b +'''</TD>
            <TD>''' + str(x[i]) + '''</TD>
          </TR>
        </TABLE>>''')
    k = []
    for q in range(len(flows) - 1):
        k.append(['action' + str(q + 1), 'action'
            + str(q)])
    print(k)
    s.edges(k)
    s.view()

# Plot final rank
def plot(a, b):
    """ a is the matrix with the flows, and b
    is a string describing the method
    """
    flows = a
    yaxes_list = [0.2] * size(flows, 0)
    plt.plot(yaxes_list, flows, 'ro')
    frame1 = plt.gca()
    frame1.axes.get_xaxis().set_visible(False)
    plt.axis([0, 0.7, min(flows) - 0.05,
        max(flows) + 0.05])
    plt.title(b + " results")
```

```
    plt.ylabel("Flows")
    plt.legend()
    plt.grid(True)
    z1 = []
    for i in range(size(flows, 0)):
        z1.append('   (Action ' + str(i + 1) + ')')
    z = [str(a) + b for a, b in zip(flows, z1)]
    for X, Y, Z in zip(yaxes_list, flows, z):
        plt.annotate('{}'.format(Z), xy = (X, Y),
            xytext = (10, -4), ha = 'left',
            textcoords = 'offset points')
    plt.show()
```

References

1. Brans, J. P., & Mareschal, B. (2005). PROMETHEE methods. In J. Figueira, S. Greco, & M. Ehrgott (Eds.), *Multiple criteria decision analysis: State of the art surveys* (pp. 163–196). New York: Springer Science + Business Media, Inc.
2. Brans, J. P., Vincke, P., & Mareschal, B. (1986). How to select and how to rank projects: The PROMETHEE method. *European Journal of Operational Research, 24*(2), 228–238.
3. Chai, J., & Liu, J. N. K. (2010). A novel multicriteria group decision making approach with intuitionistic fuzzy SIR method. In *World Automation Congress (WAC)*. Piscataway: IEEE.
4. Chan, C., Yu, K. M., & Yung K. L. (2011). Selection of solar energy for green building using superiority and inferiority multi-criteria ranking (SIR) method. In *Proceedings of the 3rd International Postgraduate Conference on Infrastructure and Environment*. Hong Kong: Hong Kong Polytechnic University
5. Chou, J. S., & Ongkowijoyo, C. S. (2015). Reliability-based decision making for selection of ready-mix concrete supply using stochastic superiority and inferiority ranking method. *Reliability Engineering & System Safety, 137*, 29–39.
6. Ma, Z. J., Zhang, N., & Dai, Y. (2014). A novel SIR method for multiple attributes group decision making problem under hesitant fuzzy environment. *Journal of Intelligent & Fuzzy Systems, 26*(5), 2119–2130.
7. Marzouk, M. (2008). A superiority and inferiority ranking model for contractor selection. *Construction Innovation, 8*(4), 250–268.
8. Marzouk, M., Shinnawy, N. E., Moselhi, O., & El-Said, M. (2013). Measuring sensitivity of procurement decisions using superiority and inferiority ranking. *International Journal of Information Technology & Decision Making, 12*(3), 395–423.
9. Rebai, A. (1993). BBTOPSIS: A bag based technique for order preference by similarity to ideal solution. *Fuzzy Sets and Systems, 60*(2), 143–162.
10. Rebai, A. (1994). Canonical fuzzy bags and bag fuzzy measures as a basis for MADM with mixed non cardinal data. *European Journal of Operational Research, 78*(1), 34–48.
11. Roy, B., Slowinski, R., & Treichel, W. (1992). Multicriteria programming of water supply systems for rural areas. *Water Resources Bulletin, 28*(1), 13–31.
12. Safari, H., Aghighi, M., Rabor, F. M., & Cruz-Machado, V. A. (2014). A new approach to job evaluation through fuzzy SIR. In *Proceedings of the 8th International Conference on Management Science and Engineering Management* (pp. 273–289). Berlin: Springer.
13. Tam, C. M., & Tong, T. K. (2008). Locating large-scale harbour-front project developments using SIR method with grey aggregation approach. *Construction Innovation, 8*(2), 120–136.
14. Xu, X. (2001). The SIR method: A superiority and inferiority ranking method for multiple criteria decision making. *European Journal of Operational Research, 131*(3), 587–602.

Chapter 5
AHP

5.1 Introduction

The analytic hierarchy process is a widely used MCDA methodology proposed by Saaty [23, 24]. AHP is based on the relative measurement theory, where we are not interested in the exact measurement of the alternatives performances over the criteria, but rather on the relative difference of one alternative over another. For instance, in the facility location problem used in previous chapters, we may not be able to evaluate the investment costs for the sites, but we may be able to make comparisons among the different sites (e.g., the investment costs in the first site are equally important to the investment costs in the second site). AHP is using pairwise comparisons between alternatives to produce a relative rating of the alternatives. It is best suited for cases where we are not interested in the precise scores of the alternatives, but on finding the best alternative. It has been successfully applied in various instances; for a comprehensive state-of-the-art literature review refer to Vaidya and Kumar [36]; Table 5.1, adopted from that reference, presents the distribution of papers on AHP by application areas.

The initial methodology was criticized by many researchers that proposed various modifications; research has focused on the ratio scales used and the method to elicit the priority vectors. Studies also emerged using fuzzy numbers [5, 7, 37] and group decision making [8, 12, 25]. This chapter will present the basic methodology proposed by Saaty [23] with comments on the various possibilities in each step; a discussion on the rank reversal phenomenon will also take place. In all cases, there will be a detailed numerical example and an implementation in Python.

Electronic Supplementary Material The online version of this chapter (https://doi.org/10.1007/978-3-319-91648-4_5) contains supplementary material, which is available to authorized users.

J. Papathanasiou, N. Ploskas, *Multiple Criteria Decision Aid*, Springer Optimization and Its Applications 136, https://doi.org/10.1007/978-3-319-91648-4_5

Table 5.1 Distribution of papers on AHP by application areas [36]

Area	N	%
Engineering	26	17.3
Personal	26	17.3
Social	23	15.3
Manufacturing	18	12
Industry	15	10
Government	13	8.7
Education	11	7.3
Political	6	4
Others	12	8
Total	150	100

5.2 Methodology

Let us assume that the decision making problem consists of m alternatives and n criteria. AHP does not use exact evaluations to form a decision matrix as in the methods presented in Chapters 1, 2, 3, and 4. Psychologists argue that it is easier and more accurate to use scales in order to evaluate an alternative over another. It is also easier for a decision maker to express an opinion on only two alternatives rather than on all the alternatives simultaneously. Therefore, AHP uses a ratio scale that does not require any units in the comparison. A judgement is a relative value of two quantities having the same units. In AHP, the decision maker does not provide a numerical judgement, but a relative verbal comparison of one alternative over another. For example, in the facility location problem used in previous chapters, the criteria weights and the decision matrix were calculated by the decision maker (Table 1.2). Instead of evaluating all alternatives with a numerical judgement over all criteria, matrices with pairwise comparisons are formed in AHP.

$$X = \begin{bmatrix} 1 & x_{12} & \cdots\cdots\cdots & x_{1n} \\ \frac{1}{x_{12}} & 1 & \cdots\cdots\cdots & \cdots \\ \cdots\cdots & \cdots & x_{ij} & \cdots\cdots \\ \cdots\cdots & \frac{1}{x_{ij}} & \cdots\cdots & \cdots \\ \cdots\cdots & \cdots\cdots & 1 & \cdots \\ \frac{1}{x_{1n}} & \cdots\cdots & \cdots\cdots & 1 \end{bmatrix}$$

Matrix X is a positive reciprocal square matrix, where x_{ij} is the comparison between element i and j. Naturally, $x_{ij} = \frac{1}{x_{ji}}$ and $x_{ji} = \frac{1}{x_{ij}}$. For instance, if site 1 has three times more employment needs than site 2, i.e., $x_{12} = 3$, then site 2 has three times less employment needs than site 1, i.e., $x_{21} = \frac{1}{3}$. In addition, $x_{ii} = 1$ since each alternative is equally important to itself. If the matrix is perfectly consistent, then the transitivity rule holds for all comparisons.

$$x_{ij} = x_{ik}x_{kj} \tag{5.1}$$

For example, if site 1 has three times more employment needs than site 2 and site 2 has two times more employment needs than site 3, then it is expected that site 1 has six ($= 2 \times 3$) times more employment needs than site 3. However, inconsistences may exist in the pairwise comparison matrices, so a consistency check should be performed (discussed later in this section).

AHP is comprised of the following seven steps:

Step 1. Form the Pairwise Comparison Matrix of the Criteria
The decision maker expresses how two criteria or alternatives compare to each other. The number of necessary comparisons for this pairwise comparison matrix is $\frac{n^2-n}{2}$. The comparisons are collected in an $n \times n$ pairwise comparison matrix. Saaty [23] proposed a 1–9 scale that is used in most AHP applications (Table 5.2). Psychologists suggest that a smaller scale would not give the same level of detail, while the decision maker would have difficulties expressing his/her opinion in a larger scale. However, many other scales have been used in AHP [16, 17, 21].

Step 2. Consistency Check on the Pairwise Comparison Matrix of the Criteria
Given the pairwise comparison matrix of the criteria, its maximum eigenvalue, λ_{max}, is equal to n if and only if the matrix is consistent (its maximum eigenvalue is greater than n if the matrix is not consistent). Saaty [23] proposed the consistency index as follows:

$$CI(X) = \frac{\lambda_{max} - n}{n - 1} \tag{5.2}$$

However, experimental results showed that the expected value of CI for a random matrix of size $n + 1$ is on average greater than the expected value of CI for a random matrix of size n. Therefore, CI is not fair in comparing matrices of

Table 5.2 The 1–9 fundamental scale [23]

Intensity of importance	Definition
1	Equal importance
2	Weak
3	Moderate importance
4	Moderate plus
5	Strong importance
6	Strong plus
7	Very strong or demonstrated importance
8	Very, very strong
9	Extreme importance

different order and needs to be rescaled. Given a pairwise comparison matrix of size n, the consistency ratio (CR), the rescaled version of CI, can be calculated as follows:

$$CR(X) = \frac{CI(X)}{RI_n} \tag{5.3}$$

where RI_n is a real number that estimates the average CI obtained from a large data set of randomly generated matrices of size n. Saaty [23] suggests that matrices with $CR \leq 0.1$ are acceptable, while matrices with $CR > 0.1$ are inconsistent. Saaty [23] calculated the random indices (RI) presented in Table 5.3. Other researchers have run simulations and calculated similar but somewhat different values [1, 20, 35].

Step 3. Compute the Priority Vector of Criteria

Several methods exist for eliciting the priority vector of criteria $w = (w_1, w_2, \cdots, w_n)$. Each method combines the pairwise comparisons into a rating. Different methods might lead to different priority vectors. We consider the following three methods:

1. **Eigenvector method**: The most widely used method to estimate a priority vector is that proposed by Saaty [23]. According to this method, a priority vector is the principal eigenvector of X. The principal eigenvector is also called the Perron–Frobenius eigenvector. Given a matrix X whose elements are obtained as ratios between weights, we multiply it by w.

$$Xw = \begin{bmatrix} \frac{w_1}{w_1} & \frac{w_1}{w_2} & \cdots & \frac{w_1}{w_n} \\ \frac{w_2}{w_1} & \frac{w_2}{w_2} & \cdots & \frac{w_2}{w_n} \\ \vdots & \cdots\cdots & & \vdots \\ \frac{w_n}{w_1} & \frac{w_n}{w_2} & \cdots & \frac{w_n}{w_n} \end{bmatrix} \begin{bmatrix} w_1 \\ w_2 \\ \vdots \\ w_n \end{bmatrix} = \begin{bmatrix} nw_1 \\ nw_2 \\ \vdots \\ nw_n \end{bmatrix} = nw$$

Table 5.3 Random indices [23]

n	RI
1	0
2	0
3	0.58
4	0.90
5	1.12
6	1.24
7	1.32
8	1.41
9	1.45
10	1.49

A formulation of the type $Xw = nw$ implies that n and w are an eigenvalue and an eigenvector of X, respectively. In addition, n is the largest eigenvalue of X. Vector w can be obtained from a pairwise comparison matrix X by solving the following equation system

$$\begin{cases} Xw = \lambda_{max} \\ w^T \left[1, 1, \cdots, 1\right]^T = 1 \end{cases} \tag{5.4}$$

It is easy to calculate the principal eigenvector of X in a software package (see Section 5.2.2), but several approximation methods also exist, e.g., the power method.

2. **The normalized column sum method**: According to this method, the priority vector is calculated as the sum of the elements on a row divided by the sum of the elements of matrix X.

$$w_i = \frac{\sum_{j=1}^{n} x_{ij}}{\sum_{i=1}^{n} \sum_{j=1}^{n} x_{ij}} \tag{5.5}$$

3. **Geometric mean method**: The geometric mean method was proposed by Crawford and Williams [11]. According to this method, the priority vector is obtained as the geometric mean of the elements on a row divided by a normalization term in order to make the sum of w equal to 1.

$$w_i = \frac{\left(\prod_{j=1}^{n} x_{ij}\right)^{\frac{1}{n}}}{\sum_{i=1}^{n} \left(\prod_{j=1}^{n} x_{ij}\right)^{\frac{1}{n}}} \tag{5.6}$$

Johnson et al. [19] presented a rank reversal problem for scale inversion with the eigenvector method. The rank reversal can be avoided by using the geometric mean method. Therefore, many researchers expressed their preference in the geometric mean method over the eigenvector method [6, 9, 15, 18]. On the other hand, Saaty's group has always supported the eigenvector method [27–30]. Various other methods have been also proposed in the literature [10, 14, 32].

Step 4. Form the Pairwise Comparison Matrices of the Alternatives for Each Criterion

Similar to Step 1, the decision maker expresses how the alternatives compare to each other for each criterion. Therefore, it creates a pairwise comparison matrix of the alternatives for each criterion. The comparisons are collected in n pairwise comparison matrices of size $m \times m$.

Step 5. Consistency Check on the Pairwise Comparison Matrices of the Alternatives

Similar to Step 2, a consistency check is performed on all pairwise comparison matrices of the alternatives. The only difference is that n (number of criteria) is replaced by m (number of alternatives) in Equations (5.2) and (5.3).

Step 6. Compute the Local Priority Vectors of the Alternatives
Similar to Step 3, the local alternative priorities are calculated for each pairwise comparison matrix of the alternatives. Any of the three methods presented in Step 3 can also be used in this step. The only difference is that n is replaced by m, and w by s_j in Equations (5.4)–(5.6), where j is the criterion to which the pairwise comparison matrix of the alternatives is associated. After this procedure is performed for all the pairwise comparison matrices of the alternatives, we form the score matrix S as follows:

$$S = \left[s_1\ s_2 \cdots s_n\right] \tag{5.7}$$

i.e., the jth column of S corresponds to the vector s_j.

Step 7. Aggregate the Local Priorities and Rank the Alternatives
In the last step, the priority criteria and local alternative priorities are combined to calculate the global alternative priorities.

$$v = Sw \tag{5.8}$$

where v is the global alternative priorities. The ith entry of v represents the global alternative priority assigned to the ith alternative. Finally, the ranking is derived by ordering the global alternative priorities in decreasing order.

5.2.1 *Numerical Example*

We use the same example that was presented in previous chapters. The main difference is that AHP does not process numerical judgement evaluations like those in Table 1.2. Therefore, we need to form the pairwise comparison matrices of the criteria and the alternatives. Initially, we start forming the pairwise comparison matrix of the criteria. Let us assume that the decision maker expressed the following comparisons among the four criteria:

- The investments costs are equally important to the employment needs ($x_{12} = x_{21} = 1$)
- The investments costs are strongly more important than the social impact ($x_{13} = 5, x_{31} = 1/5$)
- The investment costs are moderately more important than the environmental impact ($x_{14} = 3, x_{41} = 1/3$)
- The employment needs are strongly more important than the social impact ($x_{23} = 5, x_{32} = 1/5$)
- The employment needs are moderately more important than the environmental impact ($x_{24} = 3, x_{42} = 1/3$)
- The environmental impact is moderately more important than the social impact ($x_{34} = 1/3, x_{43} = 3$)

Table 5.4 Pairwise comparison matrix of the criteria

	Investment costs	Employment needs	Social impact	Environmental impact
Investment costs	1	1	5	3
Employment needs	1	1	5	3
Social impact	1/5	1/5	1	1/3
Environmental impact	1/3	1/3	3	1

Therefore, the resulting pairwise comparison matrix of the criteria is shown in Table 5.4.

Then, we perform a consistency check on the pairwise comparison matrix of the criteria. Initially, we compute the maximum eigenvalue of matrix X (using a software package); $\lambda_{max} = 4.043$. Next, we are using Equations (5.2) and (5.3) to calculate CR as follows:

$$CR(X) = \frac{CI(X)}{RI_n} = \frac{\frac{\lambda_{max}-n}{n-1}}{RI_n} = \frac{\frac{4.043-4}{4-1}}{0.90} = 0.0161$$

$CR \leq 0.1$, so the pairwise comparison matrix of the criteria is consistent.

Then, we calculate the priority vector of criteria using the eigenvector method (Equation (5.4)). The resulting vector is $w = \left[0.390\ 0.390\ 0.068\ 0.152\right]^T$.

Next, we continue forming the pairwise comparison matrices of the alternatives for each criterion. For example, the decision maker expresses the following comparisons of alternatives with regard to the criterion investments costs

- Site 1 is strongly more important than site 2 ($x_{12} = 5, x_{21} = 1/5$)
- Site 1 is equally important to site 3 ($x_{13} = x_{31} = 1$)
- Site 1 is equally important to site 4 ($x_{14} = x_{41} = 1$)
- Site 5 is moderately more important than site 1 ($x_{15} = 1/3, x_{51} = 3$)
- Site 1 is moderately more important than site 6 ($x_{16} = 3, x_{61} = 1/3$)
- ⋮
- ⋮

Similar comparisons of the alternatives for each criterion will lead us to form the pairwise comparison matrices of the alternatives for each criterion. Table 5.5 presents the pairwise comparison matrix of the alternatives with regard to investment costs, Table 5.6 presents the pairwise comparison matrix of the alternatives with regard to employment needs, Table 5.7 presents the pairwise comparison matrix of the alternatives with regard to social impact, and Table 5.8 presents the pairwise comparison matrix of the alternatives with regard to environmental impact.

Then, we perform a consistency check on the pairwise comparison matrices of the alternatives:

Table 5.5 Pairwise comparison matrix of the alternatives with regard to investment costs

	Site 1	Site 2	Site 3	Site 4	Site 5	Site 6
Site 1	1	5	1	1	1/3	3
Site 2	1/5	1	1/3	1/5	1/7	1
Site 3	1	3	1	1/3	1/5	1
Site 4	1	5	3	1	1/3	3
Site 5	3	7	5	3	1	7
Site 6	1/3	1	1	1/3	1/7	1

Table 5.6 Pairwise comparison matrix of the alternatives with regard to employment needs

	Site 1	Site 2	Site 3	Site 4	Site 5	Site 6
Site 1	1	7	3	1/3	1/3	1/3
Site 2	1/7	1	1/3	1/7	1/9	1/7
Site 3	1/3	3	1	1/5	1/5	1/5
Site 4	3	7	5	1	1	1
Site 5	3	9	5	1	1	1
Site 6	3	7	5	1	1	1

Table 5.7 Pairwise comparison matrix of the alternatives with regard to social impact

	Site 1	Site 2	Site 3	Site 4	Site 5	Site 6
Site 1	1	1/9	1/7	1/9	1	1/5
Site 2	9	1	1	1	5	3
Site 3	7	1	1	1	5	1
Site 4	9	1	1	1	7	3
Site 5	1	1/5	1/5	1/7	1	1/3
Site 6	5	1/3	1	1/3	3	1

Table 5.8 Pairwise comparison matrix of the alternatives with regard to environmental impact

	Site 1	Site 2	Site 3	Site 4	Site 5	Site 6
Site 1	1	1/5	1/5	1/3	1/7	1/5
Site 2	5	1	1	3	1/3	1
Site 3	5	1	1	1	1/3	1
Site 4	3	1/3	1	1	1/7	1
Site 5	7	3	3	7	1	5
Site 6	5	1	1	1	1/5	1

- Pairwise comparison matrix of the alternatives with regard to investment costs

$$CR(X) = \frac{CI(X)}{RI_m} = \frac{\frac{\lambda_{max}-m}{m-1}}{RI_m} = \frac{\frac{6.208-6}{6-1}}{1.24} = 0.0335$$

- Pairwise comparison matrix of the alternatives with regard to employment needs

$$CR(X) = \frac{CI(X)}{RI_m} = \frac{\frac{\lambda_{max}-m}{m-1}}{RI_m} = \frac{\frac{6.164-6}{6-1}}{1.24} = 0.0264$$

Table 5.9 Score matrix

	Investment costs	Employment needs	Social impact	Environmental impact
Site 1	0.162	0.120	0.032	0.034
Site 2	0.044	0.027	0.277	0.163
Site 3	0.098	0.054	0.222	0.133
Site 4	0.193	0.263	0.291	0.091
Site 5	0.441	0.272	0.043	0.455
Site 6	0.062	0.263	0.136	0.124

- Pairwise comparison matrix of the alternatives with regard to social impact

$$CR(X) = \frac{CI(X)}{RI_m} = \frac{\frac{\lambda_{max}-m}{m-1}}{RI_m} = \frac{\frac{6.137-6}{6-1}}{1.24} = 0.0221$$

- Pairwise comparison matrix of the alternatives with regard to environmental impact

$$CR(X) = \frac{CI(X)}{RI_m} = \frac{\frac{\lambda_{max}-m}{m-1}}{RI_m} = \frac{\frac{6.247-6}{6-1}}{1.24} = 0.0398$$

$CR \leq 0.1$ in all cases, so the pairwise comparison matrices of the alternatives are consistent.

Then, we calculate the local priority vectors using the eigenvector method (Equation (5.4)). The resulting score matrix S is presented in Table 5.9.

Finally, the global alternative priorities are calculated as follows:

$$v = Sw = \left[0.117\ 0.071\ 0.095\ 0.212\ 0.350\ 0.155\right]$$

The final ranking of the sites (from best to worst) is Site 5–Site 4–Site 6–Site 1–Site 3–Site 2.

Similarly, Tables 5.10 and 5.11 present the priority vector of criteria, the score matrix of alternatives, and the global priorities if we use the normalized column sum method and the geometric mean method, respectively. The rankings remain the same in all three methods.

5.2.2 *Python Implementation*

The file *AHP.py* includes a Python implementation of AHP. The input variables are m (the number of the alternatives, line 90), n (the number of the criteria, line 93), *RI* (the random indices for consistency checking, lines 96–97), *PCcriteria*

Table 5.10 Priority vector of criteria, score matrix of alternatives, and global priorities using the normalized column sum method

	Investment costs	Employment needs	Social impact	Environmental impact	Global priorities
w	0.389	0.389	0.069	0.154	–
Site 1	0.164	0.122	0.032	0.035	0.118
Site 2	0.045	0.028	0.274	0.162	0.072
Site 3	0.098	0.056	0.224	0.137	0.096
Site 4	0.193	0.262	0.289	0.094	0.211
Site 5	0.439	0.271	0.043	0.446	0.348
Site 6	0.062	0.262	0.137	0.127	0.155

Table 5.11 Priority vector of criteria, score matrix of alternatives, and global priorities using the geometric mean method

	Investment costs	Employment needs	Social impact	Environmental impact	Global priorities
w	0.391	0.391	0.067	0.151	–
Site 1	0.161	0.116	0.033	0.034	0.116
Site 2	0.044	0.027	0.277	0.164	0.071
Site 3	0.094	0.054	0.221	0.136	0.094
Site 4	0.194	0.264	0.293	0.090	0.212
Site 5	0.445	0.275	0.043	0.451	0.352
Site 6	0.062	0.264	0.133	0.125	0.155

(the pairwise comparison matrix of the criteria, lines 100–101), and *allPCM* (the pairwise comparison matrix of the alternatives, lines 114–139). The function *norm* calculates the priority vector of the criteria or the alternatives when the normalized column sum method is used (lines 11–19), while the function *geomean* calculates the priority vector of the criteria or the alternatives when the geometric mean method is used (lines 22–31). The function *ahp* performs all the steps of AHP and calls the other functions (lines 34–78). The eigenvector method is implemented using the function *eigs* of the *scipy* library. Consistency checks on the pairwise comparison matrix of the criteria and the pairwise comparison matrices of the alternatives are performed in the *main* function using the function *eigvals* of the *numpy* library (lines 105–111 and 143–153). The final results are displayed in line 159.

```
1.   # Filename: AHP.py
2.   # Description: Analytic hierarchy process method
3.   # Authors: Papathanasiou, J. & Ploskas, N.
4.
5.   from numpy import *
6.   import scipy.sparse.linalg as sc
7.   import matplotlib.pyplot as plt
8.   from AHP_Final_Rank_Figure import graph, plot
```

```python

# normalized column sum method
def norm(x):
   """ x is the pairwise comparison matrix for the
   criteria or the alternatives
   """
    k = array(sum(x, 0))
    z = array([[round(x[i, j] / k[j], 3)
       for j in range(x.shape[1])]
        for i in range(x.shape[0])])
    return z

# geometric mean method
def geomean(x):
    """ x is the pairwise comparison matrix for the
    criteria or the alternatives
    """
    z = [1] * x.shape[0]
    for i in range(x.shape[0]):
        for j in range(x.shape[1]):
            z[i] = z[i] * x[i][j]
        z[i] = pow(z[i], (1 / x.shape[0]))
    return z

# AHP method: it calls the other functions
def ahp(PCM, PCcriteria, m, n, c):
    """ PCM is the pairwise comparison matrix for the
    alternatives,  PCcriteria is the pairwise comparison
    matrix for the criteria, m is the number of the
    alternatives, n is the number of the criteria, and
    c is the method to estimate a priority vector (1 for
    eigenvector, 2 for normalized column sum, and 3 for
    geometric mean)
    """
    # calculate the priority vector of criteria
    if c == 1: # eigenvector
        val, vec = sc.eigs(PCcriteria, k = 1, which = 'LM')
        eigcriteria = real(vec)
        w = eigcriteria / sum(eigcriteria)
        w = array(w).ravel()
    elif c == 2: # normalized column sum
        normPCcriteria = norm(PCcriteria)
        w = array(sum(normPCcriteria, 1) / n)
    else: # geometric mean
        GMcriteria = geomean(PCcriteria)
        w = GMcriteria / sum(GMcriteria)
   # calculate the local priority vectors for the
   # alternatives
    S = []
    for i in range(n):
        if c == 1: # eigenvector
            val, vec = sc.eigs(PCM[i * m:i * m + m, 0:m],
                k = 1, which = 'LM')
            eigalter = real(vec)
```

```
            s = eigalter / sum(eigalter)
           s = array(s).ravel()
        elif c == 2: # normalized column sum
            normPCM = norm(PCM[i*m:i*m+m,0:m])
            s = array(sum(normPCM, 1) / m)
        else: # geometric mean
            GMalternatives = geomean(PCM[i*m:i*m+m,0:m])
            s = GMalternatives / sum(GMalternatives)
        S.append(s)
    S = transpose(S)

    # calculate the global priority vector for the
    # alternatives
    v = S.dot(w.T)

    return v

# main function
def main(a, b, c):
    """ a, b, and c are flags; if a and b are set to 'y'
    they do print the results, anything else does not
    print the results. If c equals 1, the eigenvector
    method is used; if c equals 2, the normalized column
    sum method is used; otherwise, the geometric mean
    method is used
    """
    # the number of the alternatives
    m = 6

    # the number of the criteria
    n = 4

    # random indices for consistency checking
    RI = [0, 0, 0.58, 0.90, 1.12, 1.24, 1.32, 1.41,
        1.45, 1.49]

    # pairwise comparison matrix of the criteria
    PCcriteria = array([[1, 1, 5, 3], [1, 1, 5, 3],
        [1/5, 1/5, 1, 1/3], [1/3, 1/3, 3, 1]])

    # consistency check for pairwise comparison matrix of
    # the criteria
    lambdamax = amax(linalg.eigvals(PCcriteria).real)
    CI = (lambdamax - n) / (n - 1)
    CR = CI / RI[n - 1]
    print("Inconsistency index of the criteria: ", CR)
    if CR > 0.1:
        print("The pairwise comparison matrix of the"
            " criteria is inconsistent")

    # pairwise comparison matrix of the alternatives
    PCM1 = array([[1, 5, 1, 1, 1/3, 3],
                 [1/5, 1, 1/3, 1/5, 1/7, 1],
                 [1, 3, 1, 1/3, 1/5, 1],
```

```
                  [1, 5, 3, 1, 1/3, 3],
                  [3, 7, 5, 3, 1, 7],
                  [1/3, 1, 1, 1/3, 1/7, 1]])
    PCM2 = array([[1, 7, 3, 1/3, 1/3, 1/3],
                  [1/7, 1, 1/3, 1/7, 1/9, 1/7],
                  [1/3, 3, 1, 1/5, 1/5, 1/5],
                  [3, 7, 5, 1, 1, 1],
                  [3, 9, 5, 1, 1, 1],
                  [3, 7, 5, 1, 1, 1]])
    PCM3 = array([[1, 1/9, 1/7, 1/9, 1, 1/5],
                  [9, 1, 1, 1, 5, 3],
                  [7, 1, 1, 1, 5, 1],
                  [9, 1, 1, 1, 7, 3],
                  [1, 1/5, 1/5, 1/7, 1, 1/3],
                  [5, 1/3, 1, 1/3, 3, 1]])
    PCM4 = array([[1, 1/5, 1/5, 1/3, 1/7, 1/5],
                  [5, 1, 1, 3, 1/3, 1],
                  [5, 1, 1, 1, 1/3, 1],
                  [3, 1/3, 1, 1, 1/7, 1],
                  [7, 3, 3, 7, 1, 5],
                  [5, 1, 1, 1, 1/5, 1]])

    allPCM = vstack((PCM1, PCM2, PCM3, PCM4))

    # consistency check for pairwise comparison matrix of
    # the alternatives
    for i in range(n):
        lambdamax = amax(linalg.eigvals(allPCM[i * m:i
           * m + m, 0:m]).real)
        CI = (lambdamax - m) / (m - 1)
        CR = CI / RI[m - 1]
        print("Inconsistency index of the alternatives for"
           " criterion ", (i + 1), ": ", CR)
        if CR > 0.1:
            print("The pairwise comparison matrix of the"
               "alternatives for criterion ", (i + 1),
               "is inconsistent")

    # call ahp method
    scores = ahp(allPCM, PCcriteria, m, n, c)

    # print results
    print("Global priorities = ", scores)

    # plot results
    if a == 'y':
        graph(around(scores, 3), "Score")
    if b == 'y':
        plot(around(scores, 3), "AHP")

if __name__ == '__main__':
    main('n', 'y', 1)
```

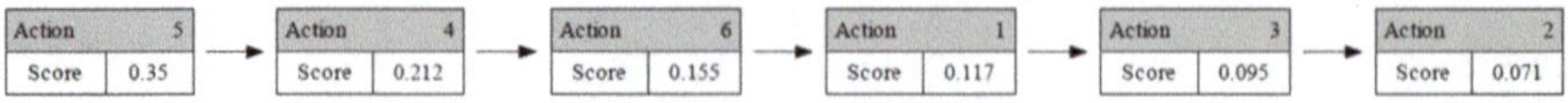

Fig. 5.1 Final rank figure plotted with graphviz module

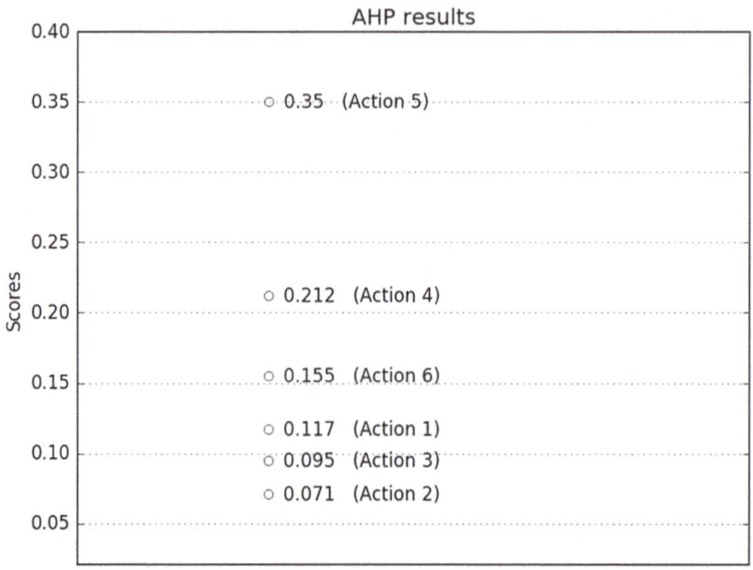

Fig. 5.2 Final rank plotted with matplotlib

The file *AHP_Final_Rank_Figure.py* is an optional module that includes two methods to plot the results of AHP. This module produces two plots with the results (Figures 5.1 and 5.2 display the results when the eigenvector method is used). In order to run these methods, one needs to install graphviz and matplotlib modules.

```
# Filename: AHP_Final_Rank_Figure.py
# Description: Optional module to plot the
# results of AHP method
# Authors: Papathanasiou, J. & Ploskas, N.

import matplotlib.pyplot as plt
from graphviz import Digraph
from numpy import *

# Plot final rank figure
def graph(scores, b):
    """ scores is the matrix with the scores, and b
    is a string describing the score
    """
    s = Digraph('Actions', node_attr = {'shape':
        'plaintext'})
    s.body.extend(['rankdir = LR'])
```

```
    x = sort(scores)
    y = argsort(scores)
    l = []
    for i in y:
        s.node('action' + str(i), '''<
        <TABLE BORDER="0" CELLBORDER="1"
            CELLSPACING="0" CELLPADDING="4">
          <TR>
            <TD COLSPAN="2" bgcolor="grey" >Action
                ''' + str(y[i] + 1) + '''</TD>
          </TR>
          <TR>
            <TD>'''+ b +'''</TD>
            <TD>''' + str(x[i]) + '''</TD>
          </TR>
        </TABLE>>''')
    k = []
    for q in range(len(scores) - 1):
        k.append(['action' + str(q + 1), 'action'
            + str(q)])
    print(k)
    s.edges(k)
    s.view()

# Plot final rank
def plot(a, b):
    """ a is the matrix with the scores, and b
    is a string describing the method
    """
    scores = a
    yaxes_list = [0.2] * size(scores, 0)
    plt.plot(yaxes_list, scores, 'ro')
    frame1 = plt.gca()
    frame1.axes.get_xaxis().set_visible(False)
    plt.axis([0, 0.7, min(scores) - 0.05,
        max(scores) + 0.05])
    plt.title(b + " results")
    plt.ylabel("Scores")
    plt.legend()
    plt.grid(True)
    z1 = []
    for i in range(size(scores, 0)):
        z1.append('    (Action ' + str(i + 1) + ')')
    z = [str(a) + b for a, b in zip(scores, z1)]
    for X, Y, Z in zip(yaxes_list, scores, z):
        plt.annotate('{}'.format(Z), xy = (X, Y),
            xytext=(10, -4), ha = 'left',
            textcoords = 'offset points')
    plt.show()
```

5.2.3 Rank Reversal

Similar to TOPSIS (Section 1.2.3), AHP is also not immune to the rank reversal phenomenon. In the next part of this section, we will recreate the example presented by Belton and Gear [4] to demonstrate the rank reversal phenomenon in AHP.

They consider three different alternatives, A, B, and C, and three criteria, a, b, and c. They assume that all criteria are equally important; therefore, the pairwise comparison matrix of the criteria is shown in Table 5.12.

Let us perform a consistency check on the pairwise comparison matrix of the criteria. Initially, we compute the maximum eigenvalue of the pairwise comparison matrix (using a software package); $\lambda_{max} = 3$. Next, we are using Equations (5.2) and (5.3) to calculate CR as follows:

$$CR(X) = \frac{CI(X)}{RI_n} = \frac{\frac{\lambda_{max}-n}{n-1}}{RI_n} = \frac{\frac{3-3}{3-1}}{0.58} = 0$$

$CR \leq 0.1$, so the pairwise comparison matrix of the criteria is consistent. Of course, this result was expected since all criteria are equally important, i.e., we have a matrix of ones.

Then, we calculate the priority vector of criteria using the eigenvector method (Equation (5.4)). As expected, the resulting vector is $w = \left[1/3\ 1/3\ 1/3\right]^T$.

The pairwise comparison matrices of the alternatives used by Belton and Gear [4] are shown in Tables 5.13, 5.14, and 5.15. Table 5.13 presents the pairwise comparison matrix of the alternatives with regard to the criterion a, Table 5.14 presents the pairwise comparison matrix of the alternatives with regard to the criterion b, and Table 5.15 presents the pairwise comparison matrix of the alternatives with regard to the criterion c.

Table 5.12 Pairwise comparison matrix of the criteria

	a	b	c
a	1	1	1
b	1	1	1
c	1	1	1

Table 5.13 Pairwise comparison matrix of the alternatives with regard to the criterion a

	A	B	C
A	1	1/9	1
B	9	1	9
C	1	1/9	1

Table 5.14 Pairwise comparison matrix of the alternatives with regard to the criterion b

	A	B	C
A	1	9	9
B	1/9	1	1
C	1/9	1	1

Table 5.15 Pairwise comparison matrix of the alternatives with regard to the criterion c

	A	B	C
A	1	8/9	8
B	9/8	1	9
C	1/8	1/9	1

Table 5.16 Score matrix

	a	b	c
A	0.091	0.818	0.444
B	0.818	0.091	0.500
C	0.091	0.091	0.056

Then, we perform a consistency check on the pairwise comparison matrices of the alternatives:

- Pairwise comparison matrix of the alternatives with regard to criterion a

$$CR(X) = \frac{CI(X)}{RI_m} = \frac{\frac{\lambda_{max}-m}{m-1}}{RI_m} = \frac{\frac{3-3}{3-1}}{0.58} = 0$$

- Pairwise comparison matrix of the alternatives with regard to criterion b

$$CR(X) = \frac{CI(X)}{RI_m} = \frac{\frac{\lambda_{max}-m}{m-1}}{RI_m} = \frac{\frac{3-3}{3-1}}{0.58} = 0$$

- Pairwise comparison matrix of the alternatives with regard to criterion c

$$CR(X) = \frac{CI(X)}{RI_m} = \frac{\frac{\lambda_{max}-m}{m-1}}{RI_m} = \frac{\frac{3-3}{3-1}}{0.58} = 0$$

$CR \leq 0.1$ in all cases, so the pairwise comparison matrices of the alternatives are consistent.

Then, we calculate the local priority vectors using the eigenvector method (Equation (5.4)). The resulting score matrix S is presented in Table 5.16.

Finally, the global alternative priorities are calculated as follows:

$$v = Sw = \left[0.451\ 0.470\ 0.079\right]$$

The final ranking of the alternatives (from best to worst) is B–A–C.

Now, let us assume that a new alternative is added in this decision making problem. The new pairwise comparison matrices of the alternatives used by Belton and Gear [4] are shown in Tables 5.17, 5.18, and 5.19. Table 5.17 presents the pairwise comparison matrix of the alternatives with regard to the criterion a, Table 5.18 presents the pairwise comparison matrix of the alternatives with regard to the criterion b, and Table 5.19 presents the pairwise comparison matrix of the alternatives with regard to the criterion c.

Table 5.17 New pairwise comparison matrix of the alternatives with regard to the criterion *a*

	A	*B*	*C*	*D*
A	1	1/9	1	1/9
B	9	1	9	1
C	1	1/9	1	1/9
D	9	1	9	1

Table 5.18 New pairwise comparison matrix of the alternatives with regard to the criterion *b*

	A	*B*	*C*	*D*
A	1	9	9	9
B	1/9	1	1	1
C	1/9	1	1	1
D	1/9	1	1	1

Table 5.19 New pairwise comparison matrix of the alternatives with regard to the criterion *c*

	A	*B*	*C*	*D*
A	1	8/9	8	8/9
B	9/8	1	9	1
C	1/8	1/9	1	1/9
D	9/8	1	9	1

Then, we perform a consistency check on the pairwise comparison matrices of the alternatives

- Pairwise comparison matrix of the alternatives with regard to criterion *a*

$$CR(X) = \frac{CI(X)}{RI_m} = \frac{\frac{\lambda_{max}-m}{m-1}}{RI_m} = \frac{\frac{4-4}{4-1}}{0.90} = 0$$

- Pairwise comparison matrix of the alternatives with regard to criterion *b*

$$CR(X) = \frac{CI(X)}{RI_m} = \frac{\frac{\lambda_{max}-m}{m-1}}{RI_m} = \frac{\frac{4-4}{4-1}}{0.90} = 0$$

- Pairwise comparison matrix of the alternatives with regard to criterion *c*

$$CR(X) = \frac{CI(X)}{RI_m} = \frac{\frac{\lambda_{max}-m}{m-1}}{RI_m} = \frac{\frac{4-4}{4-1}}{0.90} = 0$$

$CR \leq 0.1$ in all cases, so the pairwise comparison matrices of the alternatives are consistent.

Then, we calculate the local priority vectors using the eigenvector method (Equation (5.4)). The resulting score matrix S is presented in Table 5.20.

Finally, the global alternative priorities are calculated as follows:

$$v = Sw = \left[0.365\ 0.289\ 0.057\ 0.289\right]$$

Table 5.20 Score matrix

	a	b	c
A	0.050	0.750	0.300
B	0.450	0.083	0.333
C	0.050	0.083	0.037
D	0.450	0.083	0.333

The final ranking of the alternatives (from best to worst) is A–B–D–C. We observe that the ranking has changed. In the initial example, B was preferred over A, but now A is preferred over B. There was no change in the relative preferences of A over B between the two examples, so the fact that the overall preference does not remain unchanged causes the rank reversal phenomenon.

The rank reversal phenomenon has led to a heated debate about the validity of AHP [13, 16, 22, 26, 33]. In order to avoid the rank reversal in AHP, Belton and Gear [4] proposed the normalization of the eigenvector weights of the alternatives using their maximum value rather than their sum. This method is called B-G modified AHP. However, Saaty and Vargas [31] presented an example where the B-G modified AHP was also subject to the rank reversal phenomenon. Schoner and Wedley [33] proposed a modified AHP method, called referenced AHP, to avoid rank reversal. The referenced AHP method requires the modification of the criteria weights when an alternative is added or deleted. Later, Schoner et al. [34] presented a normalization method and a linking pin AHP, where one alternative is chosen for each criterion as the link for criteria comparisons. Barzilai and Golany [2] showed that no normalization method can prevent rank reversal. They also proposed a multiplicative aggregation rule that replaces normalized weight vectors with weight-ratio matrices. Later, Barzilai and Lootsma [3] proposed the multiplicative AHP method for avoiding the rank reversal. However, Vargas [38] presented an example showing that the multiplicative AHP is invalid.

References

1. Alonso, J. A., & Lamata, M. T. (2006). Consistency in the analytic hierarchy process: A new approach. *International Journal of Uncertainty, Fuzziness and Knowledge-Based Systems, 14*(04), 445–459.
2. Barzilai, J., & Golany, B. (1994). AHP rank reversal, normalization and aggregation rules. *Information Systems and Operational Research, 32*(2), 57–64.
3. Barzilai, J., & Lootsma, F. A. (1997). Power relations and group aggregation in the multiplicative AHP and SMART. *Journal of Multi-Criteria Decision Analysis, 6*(3), 155–165.
4. Belton, V., & Gear, T. (1983). On a short-coming of Saaty's method of analytic hierarchies. *Omega*, 11(3), 228–230.
5. Buckley, J. J. (1985). Fuzzy hierarchical analysis. *Fuzzy Sets and Systems, 17*(3), 233–247.
6. Budescu, D. V., Zwick, R., & Rapoport, A. (1986). A comparison of the eigenvalue method and the geometric mean procedure for ratio scaling. *Applied Psychological Measurement, 10*(1), 69–78.

7. Chang, D. Y. (1996). Applications of the extent analysis method on fuzzy AHP. *European Journal of Operational Research, 95*(3), 649–655.
8. Chen, C. T. (2000). Extensions of the TOPSIS for group decision-making under fuzzy environment. *Fuzzy Sets and Systems, 114*(1), 1–9.
9. Choo, E. U., & Wedley, W. C. (2004). A common framework for deriving preference values from pairwise comparison matrices. *Computers & Operations Research*, 31(6), 893–908.
10. Cook, W. D., & Kress, M. (1988). Deriving weights from pairwise comparison ratio matrices: An axiomatic approach. *European Journal of Operational Research, 37*(3), 355–362.
11. Crawford, G., & Williams, C. (1985). A note on the analysis of subjective judgement matrices. *Journal of Mathematical Psychology, 29*(4), 387–405.
12. Dong, Y., Zhang, G., Hong, W. C., & Xu, Y. (2010). Consensus models for AHP group decision making under row geometric mean prioritization method. *Decision Support Systems, 49*(3), 281–289.
13. Dyer, J. S. (1990). Remarks on the analytic hierarchy process. *Management Science, 36*(3), 249–258.
14. Fichtner, J. (1986). On deriving priority vectors from matrices of pairwise comparisons. *Socio-Economic Planning Sciences, 20*(6), 341–345.
15. Golany, B., & Kress, M. (1993). A multicriteria evaluation of methods for obtaining weights from ratio-scale matrices. *European Journal of Operational Research, 69*(2), 210–220.
16. Harker, P. T., & Vargas, L. G. (1987). The theory of ratio scale estimation: Saaty's analytic hierarchy process. *Management Science, 33*(11), 1383–1403.
17. Ishizaka, A., Balkenborg, D., & Kaplan, T. (2011). Influence of aggregation and measurement scale on ranking a compromise alternative in AHP. *Journal of the Operational Research Society, 62*(4), 700–710.
18. Ishizaka, A., & Lusti, M. (2004). An expert module to improve the consistency of AHP matrices. *International Transactions in Operational Research, 11*(1), 97–105.
19. Johnson, C. R., Beine, W. B., & Wang, T. J. (1979). Right-left asymmetry in an eigenvector ranking procedure. *Journal of Mathematical Psychology, 19*(1), 61–64.
20. Lane, E. F., & Verdini, W. A. (1989). A consistency test for AHP decision makers. *Decision Sciences, 20*(3), 575–590.
21. Lootsma, F. A. (1989). Conflict resolution via pairwise comparison of concessions. *European Journal of Operational Research, 40*(1), 109–116.
22. Millet, I., & Saaty, T. L. (2000). On the relativity of relative measures - accommodating both rank preservation and rank reversals in the AHP. *European Journal of Operational Research, 121*(1), 205–212.
23. Saaty, T. (1977). A scaling method for priorities in hierarchical structures. *Journal of Mathematical Psychology, 15*(3), 234–281.
24. Saaty, T. (1980). *The analytic hierarchy process*. New York: McGraw-Hill.
25. Saaty, T. (1989). *The analytic hierarchy process*. Berlin: Springer.
26. Saaty, T. (1990). An exposition of the AHP in reply to the paper "remarks on the analytic hierarchy process". *Management Science, 36*(3), 259–268.
27. Saaty, T. (2003). Decision-making with the AHP: Why is the principal eigenvector necessary. *European Journal of Operational Research, 145*(1), 85–91.
28. Saaty, T., & Hu, G. (1998). Ranking by eigenvector versus other methods in the analytic hierarchy process. *Applied Mathematics Letters, 11*(4), 121–125.
29. Saaty, T., & Vargas, L. G. (1984). Comparison of eigenvalue, logarithmic least squares and least squares methods in estimating ratios. *Mathematical Modelling, 5*(5), 309–324.
30. Saaty, T., & Vargas, L. G. (1984). Inconsistency and rank preservation. *Journal of Mathematical Psychology, 28*(2), 205–214.
31. Saaty, T., & Vargas, L. G. (1984). The legitimacy of rank reversal. *Omega, 12*(5), 513–516.
32. Salo, A. A., & Hämäläinen, R. P. (1997). On the measurement of preferences in the analytic hierarchy process. *Journal of Multi-Criteria Decision Analysis, 6*(6), 309–319.
33. Schoner, B., & Wedley, W. C. (1989). Ambiguous criteria weights in AHP: Consequences and solutions. *Decision Sciences, 20*(3), 462–475.

34. Schoner, B., Wedley, W. C., & Choo, E. U. (1993). A unified approach to AHP with linking pins. *European Journal of Operational Research, 64*(3), 384–392.
35. Tummala, V. R., & Wan, Y. W. (1994). On the mean random inconsistency index of analytic hierarchy process (AHP). *Computers & Industrial Engineering, 27*(1–4), 401–404.
36. Vaidya, O. S., & Kumar, S. (2006). Analytic hierarchy process: An overview of applications. *European Journal of Operational Research, 169*(1), 1–29.
37. Van Laarhoven, P. J. M., & Pedrycz, W. (1983). A fuzzy extension of Saaty's priority theory. *Fuzzy Sets and Systems, 11*(1–3), 229–241.
38. Vargas, L. G. (1997). Comments on Barzilai and Lootsma: Why the multiplicative AHP is invalid: A practical counterexample. *Journal of Multi-Criteria Decision Analysis, 6*(3), 169–170.

Chapter 6
Goal Programming

6.1 Introduction

Goal Programming can be considered as an extension of linear programming to handle multiple goals. The roots of Goal Programming lie in the paper by Charnes et al. [3] in which they deal with executive compensation methods. Later, Charnes and Cooper [2] introduced the term Goal Programming and provided a more formal theory about it. The seminal works by Ijiri [11], Lee [14], and Ignizio [9] led Goal Programming to be one of the most widely used MCDA method. It has been applied successfully in various application areas; Table 6.1 is adopted from [19] and includes the applications of Goal Programming according to the survey conducted by the author of that book. Application domains include accounting [16], agriculture [15], economics [5], engineering [10], finance [8], government [4], international context [20], management [13], and marketing [1].

There are many variants of Goal Programming. This chapter will initially present the classical Goal Programming. In addition, we will also present the three most widely used variants: (1) Weighted Goal Programming, (2) Lexicographic Goal Programming, and (3) Chebyshev Goal Programming. In all cases, there will be a detailed numerical example and an implementation in Python.

Electronic Supplementary Material The online version of this chapter (https://doi.org/10.1007/978-3-319-91648-4_6) contains supplementary material, which is available to authorized users.

J. Papathanasiou, N. Ploskas, *Multiple Criteria Decision Aid*, Springer Optimization and Its Applications 136, https://doi.org/10.1007/978-3-319-91648-4_6

Table 6.1 Distribution of papers on Goal Programming by application areas [19]

Area	N	%
Accounting	38	5.1
Agriculture	63	8.5
Economics	28	3.7
Engineering	25	3.4
Finance	112	15.0
Government	169	22.7
International context	42	5.6
Management	244	32.8
Marketing	24	3.2
Total	745	100

6.2 Classical Goal Programming

A classical Goal Programming problem has n decision variables $x = x_1, x_2, \cdots, x_n$, and m goals. Each goal has an achieved value, $f_i(x)$, where $i = 1, 2, \cdots, m$, on its associated criterion. This value is a function of the decision variables. The decision maker sets a target value, b_i for each goal. Hence, each goal is formulated as

$$f_i(x) + n_i - p_i = b_i$$

where n_i is the negative deviational variable and p_i is the positive deviational variable of the ith goal. A deviational variable measures the difference between the target level of a goal and the value that is actually achieved in a given solution. If the achieved value is below the target level of a goal, then the difference is given by the value of the negative deviational variable, n. If the achieved value is above the target level of a goal, then the difference is given by the value of the positive deviational variable, p. Hence, a negative deviational variable shows the level by which a target level is under-achieved, while a positive deviational variable shows the level by which the target level is over-achieved. All deviational variables take non-negative values. At least one of the positive and negative deviational variables of a goal is equal to 0 in a given solution. The target values, b_i, should be set at appropriate levels. If they are pessimistic, then the resulting solution may be Pareto inefficient [18], and a Pareto efficiency restoration technique [21] will need to be employed in order to restore Pareto efficiency to the solution (for an overview of these techniques, see [12]). On the other hand, if they are optimistic, then the problem of redundancy will occur [18], in which only a few goals are taken into consideration. Hence, the decision makers should be careful when selecting the target values of each goal.

For each goal, the decision maker must decide the deviational variable(s) that will be penalized. This decision depends on the type of the goal. There are three types of goals:

- Goals that the decision maker does not want to be over-achieved. These goals are the equivalent "≤" inequality constraints in linear programming problems. For example, a goal that involves cost lies in this category. In these goals, the positive deviational variable, p, is penalized in the objective function.
- Goals that the decision maker does not want to be under-achieved. These goals are the equivalent "≥" inequality constraints in linear programming problems. For example, a goal that involves profit lies in this category. In these goals, the negative deviational variable, n, is penalized in the objective function.
- Goals that the decision maker does not want to be either under-achieved or over-achieved. These goals are the equivalent equality constraints in linear programming problems. For example, a goal that involves employment lies in this category. In these goals, both the negative and the positive deviational variables are penalized in the objective function.

After finding out the type of each goal, the objective function is formulated as

$$\min \quad h(n, p)$$

where n is the vector of m negative deviational variables, and p is the vector of m positive deviational variables. The objective function shows the distance from the target to the achieved level of the goals.

There are two types of constraints in a Goal Programming problem: (1) soft constraints, and (2) hard constraints. A soft constraint represents a goal, in which we add deviational variables to measure the difference between the target and the actual value in a given solution. A hard constraint is a constraint that should be satisfied in order for a solution to be feasible. Hence, we should determine which constraints are soft, i.e., they are goals that we would like to achieve, and which constraints are hard, i.e., we should meet them in order to find a feasible solution. Hard constraints are formulated as

$$x \in F$$

where F is the feasible region that satisfies all hard constraints. F also satisfies bound constraints. A bound constraint limits a decision or deviational variable to take certain values within its range. For example, most decision and deviational variables are set to be non-negative and continuous.

To sum up, a Goal Programming problem is formulated using the following steps:

1. Determine whether a constraint is soft or hard.
2. Add a negative and a positive deviational variable on each constraint. Determine the type of the constraint and add to the objective function the deviational variable(s) to be penalized.
3. Each hard constraint is written as a typical linear programming constraint.
4. Add bound constraints to the problem (if applicable).

A generic form of a Goal Programming problem is the following

$$\begin{aligned} \min z &= h(n, p) \\ \text{s.t.} \quad & f_i(x) + n_i + p_i = b_i \\ & x \in F \\ n_i, p_i &\geq 0, i = 1, 2, \ldots, m \end{aligned}$$

After formulating a Goal Programming problem, one can use a linear programming solver to solve it and find the values of the decision variables x and the values of the deviational variables n and p. In this chapter, we are using Pyomo [7, 17] to solve Goal Programming problems. Pyomo is an open source collection of Python software packages for formulating optimization models. The main advantage of Pyomo is that it allows formulating optimization problems in a manner that is similar to the notation commonly used in mathematical optimization. Pyomo supports numerous solvers, both open source and commercial.

We will demonstrate the basic usage of Pyomo by solving the two simple examples that were also solved with Pulp in Chapter 3. Initially, we will solve the following simple linear programming problem

$$\begin{aligned} \min z &= 60x_1 + 40x_2 \\ \text{s.t.} \quad & 4x_1 + 4x_2 \geq 10 \\ & 2x_1 + x_2 \geq 4 \\ & 6x_1 + 2x_2 \leq 12 \\ x_1 &\geq 0, x_2 \geq 0 \end{aligned}$$

The file *PYOMO_example_1.py* includes a Python implementation for solving this linear programming problem with Pyomo. Since it has only a couple of decision variables the results can be plotted as shown in Figure 6.1. This figure includes an

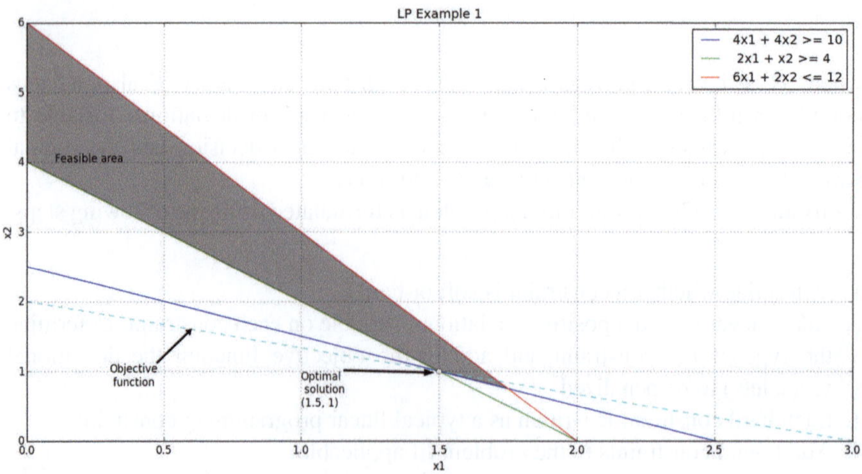

Fig. 6.1 Graphical solution of the first Pyomo example

abundance of information; the feasible region (the grey shaded area), the constraints line segments, the model constraints, the optimal solution, and the line vertical to the objective function vector through the corner with the optimal objective function value found.

Initially, an object to perform optimization is created (line 12). Any linear programming solver can be used instead of CPLEX. Next, an object of a concrete model is created (line 15). Then, we define the decision variables (lines 18–19) as non-negative continuous variables. Next, we define the objective function (lines 22–23) and the constraints (lines 26–31) of the problem. Finally, we solve the linear programming problem (line 34), then print (lines 37–44) and plot (lines 47–73) the results.

```
1.    # Filename: PYOMO_example_1.py
2.    # Description: An example of solving a linear
3.    # programming problem with Pyomo
4.    # Authors: Papathanasiou, J. & Ploskas, N.
5.
6.    from pyomo.environ import *
7.    from pyomo.opt import SolverFactory
8.    import matplotlib.pyplot as plt
9.    import numpy as np
10.
11.   # Create an object to perform optimization
12.   opt = SolverFactory('cplex')
13.
14.   # Create an object of a concrete model
15.   model = ConcreteModel()
16.
17.   # Define the decision variables
18.   model.x1 = Var(within=NonNegativeReals)
19.   model.x2 = Var(within=NonNegativeReals)
20.
21.   # Define the objective function
22.   model.obj = Objective(expr = 60 * model.x1 +
23.       40 * model.x2)
24.
25.   # Define the constraints
26.   model.con1 = Constraint(expr = 4 * model.x1 +
27.       4 * model.x2 >= 10)
28.   model.con2 = Constraint(expr = 2 * model.x1 +
29.       model.x2 >= 4)
30.   model.con3 = Constraint(expr = 6 * model.x1 +
31.       2 * model.x2 <= 12)
32.
33.   # Solve the linear programming problem
34.   results = opt.solve(model)
35.
36.   # Print the results
37.   print ("Status: ",
38.       results.solver.termination_condition)
39.
```

```
print("x1 = ", model.x1.value)
print("x2 = ", model.x2.value)

print ("The optimal value of the objective function "
    "is = ", model.obj())

# Plot the results
x = np.arange(0, 5)

plt.plot(x, 2.5 - x, label = '4x1 + 4x2 >= 10')
plt.plot(x, 4 - 2 * x, label= ' 2x1 + x2 >= 4')
plt.plot(x, 6 - 3 * x, label = '6x1 + 2x2 <= 12')
plt.plot(x, 2 - 2/3*x, '--')
plt.annotate('Objective\n function', xy = (0.6, 1.6),
    xytext = (0.3, 0.8), arrowprops =
    dict(facecolor = 'black', width = 1.5, headwidth = 7))
plt.annotate('Optimal\n solution\n(1.5, 1)',
    xy = (1.48, 0.98), xytext = (1, 0.5), arrowprops =
    dict(facecolor = 'black', width = 1.5, headwidth = 7))
plt.plot(1.5, 1, 'wo')
plt.text(0.1, 4, 'Feasible area', size = '12')

# Define the boundaries of the feasible area in the plot
a = [0, 1.5, 1.75, 0, 0]
b = [4, 1, 0.75, 6, 4]
plt.fill(a, b, 'grey')

plt.xlabel("x1")
plt.ylabel("x2")
plt.title('LP Example 1')
plt.axis([0, 3, 0, 6])
plt.grid(True)
plt.legend()
plt.show()
```

The output of the code in this case (the solution) is as follows:

```
Status: optimal
x1 = 1.5
x2 = 1.0
The optimal value of the objective function is = 130.0
```

Pyomo also solves mixed integer and binary linear programming problems (among other types of optimization problems). We will solve the following binary linear programming problem

$$
\begin{array}{llllllll}
\min z = & x_1 & + x_2 & + x_3 & + x_4 & + x_5 & + x_6 & \\
\text{s.t.} & x_1 & & + x_3 & & & & \geq 2 \\
 & x_1 & + x_2 & & & + x_5 & & \geq 2 \\
 & & & x_3 & + x_4 & & & \geq 1 \\
 & x_1 & & & + x_4 & + x_5 & & \geq 1 \\
 & & & & x_4 & + x_5 & + x_6 & \geq 1 \\
\multicolumn{8}{l}{x_i \in \{0, 1\}, i = 1, 2, \cdots, 6}
\end{array}
$$

The file *PYOMO_example_2.py* includes a Python implementation for solving this linear programming problem with Pyomo. The only difference from the previous example is that now the programmer needs to explicitly declare that he/she will be using binary variables in the variable definition function (lines 18–23). Initially, an object to perform optimization is created (line 12). Any integer linear programming solver can be used instead of CPLEX. Next, an object of a concrete model is created (line 15). Then, we define the decision variables (lines 18–23) as binary variables. Next, we define the objective function (lines 26–28) and the constraints (lines 31–40) of the problem. Finally, we solve the binary linear programming problem (line 43), and print the results (lines 46–57).

```
1.   # Filename: PYOMO_example_2.py
2.   # Description: An example of solving a binary
3.   # linear programming problem with Pyomo
4.   # Authors: Papathanasiou, J. & Ploskas, N.
5.
6.   from pyomo.environ import *
7.   from pyomo.opt import SolverFactory
8.   import matplotlib.pyplot as plt
9.   import numpy as np
10.
11.  # Create an object to perform optimization
12.  opt = SolverFactory('cplex')
13.
14.  # Create an object of a concrete model
15.  model = ConcreteModel()
16.
17.  # Define the decision variables
18.  model.x1 = Var(within=Binary)
19.  model.x2 = Var(within=Binary)
20.  model.x3 = Var(within=Binary)
21.  model.x4 = Var(within=Binary)
22.  model.x5 = Var(within=Binary)
23.  model.x6 = Var(within=Binary)
24.
25.  # Define the objective function
26.  model.obj = Objective(expr = model.x1 +
27.      model.x2 + model.x3 + model.x4 + model.x5 +
28.      model.x6)
29.
30.  # Define the constraints
31.  model.con1 = Constraint(expr = model.x1 +
32.      model.x3 >= 2)
33.  model.con2 = Constraint(expr = model.x1 +
34.      model.x2 + model.x5 >= 2)
35.  model.con3 = Constraint(expr = model.x3 +
36.      model.x4 >= 1)
37.  model.con4 = Constraint(expr = model.x1 +
38.      model.x4 + model.x5 >= 1)
39.  model.con5 = Constraint(expr = model.x4 +
40.      model.x5 + model.x6 >= 1)
```

```

# Solve the binary linear programming problem
results = opt.solve(model)

# Print the results
print ("Status: ",
    results.solver.termination_condition)

print("x1 = ", model.x1.value)
print("x2 = ", model.x2.value)
print("x3 = ", model.x3.value)
print("x4 = ", model.x4.value)
print("x5 = ", model.x5.value)
print("x6 = ", model.x6.value)

print ("The optimal value of the objective function "
    "is = ", model.obj())
```

The output of the code in this case (the solution) is as follows:

```
Status: optimal
x1 = 1.0
x2 = 0.0
x3 = 1.0
x4 = 0.0
x5 = 1.0
x6 = 0.0
The optimal value of the objective function is = 3.0
```

6.2.1 *Numerical Example*

The following example will be used throughout this chapter. A store produces and sells shirts and jackets. The price of a shirt is at 100 €, while the price of a jacket is at 90 €. Every shirt needs 2 m^2 of cotton and 5 m^2 of linen, while every jacket needs 4 m^2 of cotton and 3 m^2 of linen. The store can buy from its supplier 600 m^2 of cotton and 700 m^2 of linen every week. The company aims to achieve a weekly profit of 18,000 €. The production time of a pair of each product is 2 man-hours. The company employs 10 people in the production department and would like to keep within the 380 available hours of work each week. The manufacturing capacity of the machine for all products combined is limited to a maximum of 200 products per week. The company has signed a contract with a customer to provide his/her company with at least 60 units of each product per week.

In this problem, there are two unknown variables, the number of shirts and the number of jackets that the store should produce and sell each week. Those two variables are called the decision variables. Let x_1 be the number of shirts and x_2 the number of jackets.

If we treat the problem as a classical linear programming problem, we can identify the following constraints

1. The store has only available 600 m^2 of cotton every week and every shirt needs 2 m^2 of cotton, while every jacket needs 4 m^2 of cotton. Hence, we can derive the following constraint

$$2x_1 + 4x_2 \leq 600 \text{ (the constraint for the available cotton)}$$

2. Similarly, the store has only available 700 m^2 of linen every week and every shirt needs 5 m^2 of linen, while every jacket needs 3 m^2 of linen. Hence, we can derive the following constraint

$$5x_1 + 3x_2 \leq 700 \text{ (the constraint for the available linen)}$$

3. The company aims to achieve a weekly profit of 18,000 € and the price of a shirt is at 100 €, while the price of a jacket is at 90 €. Hence, we can derive the following constraint

$$100x_1 + 90x_2 \geq 18{,}000 \text{ (the constraint for the profit)}$$

4. The company would like to keep within the 380 available hours of work each week and the production time of a pair of each product is 2 man-hours. Hence, we can derive the following constraint

$$2x_1 + 2x_2 \leq 380 \text{ (the constraint for the production)}$$

5. The manufacturing capacity of the machine for all products combined is limited to a maximum of 200 products per week. Hence, we can derive the following constraint

$$x_1 + x_2 \leq 200 \text{ (the constraint for the manufacturing)}$$

6. The company has signed a contract with a customer to provide his/her company with at least 60 units of each product per week. Hence, we can derive the following two constraints

$$x_1 \geq 60 \text{ (the constraint for the shirts produced)}$$

$$x_2 \geq 60 \text{ (the constraint for the jackets produced)}$$

If we draw the previously mentioned constraints (Figure 6.2), we will find out that this linear programming problem is infeasible. The arrows near each constraint in Figure 6.2 show the direction in which its constraint is satisfied. Some of the conflicts can be seen easily. For example, Figure 6.3 includes only the first three constraints. The feasible regions in Figure 6.3 do not take into account that the

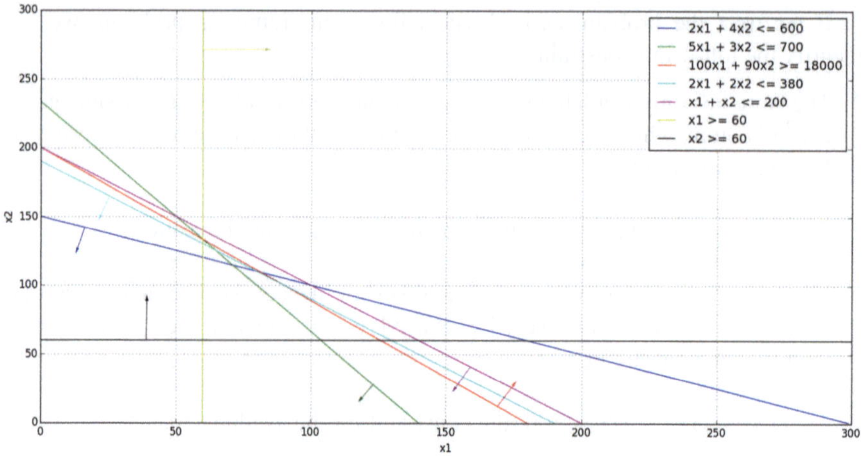

Fig. 6.2 Infeasible linear programming problem

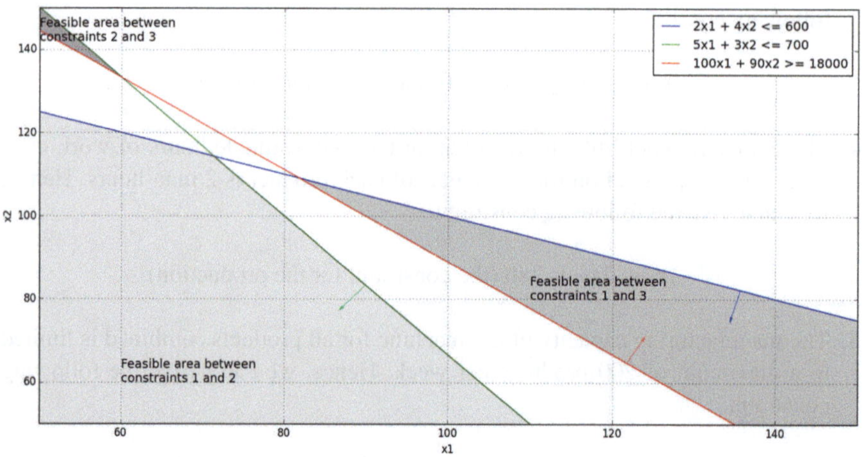

Fig. 6.3 An example of conflicting constraints

variables are integers. We observe that each pair of constraints has a feasible region but when we consider all three constraints together, the linear programming problem is infeasible.

Therefore, we will follow the steps described in the previous section to formulate the problem as a Goal Programming problem. Initially, we want to determine whether a constraint is soft or hard. For each soft constraint (goal), we will add a negative and a positive deviational variable and add to the objective function the deviational variable(s) to be penalized.

The constraint for the available cotton is a soft constraint since the company can order more cotton. The first goal can be written as

$$2x_1 + 4x_2 + n_1 - p_1 = 600$$

The variable p_1 is the unwanted deviation away from 600 m^2. Hence, we will penalize it in the objective function.

Similarly, the constraint for the available linen is a soft constraint since the company can order more linen. The second goal can be written as

$$5x_1 + 3x_2 + n_2 - p_2 = 700$$

The variable p_2 is the unwanted deviation away from 700 m^2. Hence, we will penalize it in the objective function.

The constraint for the profit is a soft constraint since the company wishes to achieve a weekly profit of 18,000 € only if this is possible. The third goal can be written as

$$100x_1 + 90x_2 + n_3 - p_3 = 18{,}000$$

The variable n_3 is the unwanted deviation away from 18,000 €. Hence, we will penalize it in the objective function.

The constraint for the production is a soft constraint since the employees can work overtime if needed. The fourth goal can be written as

$$2x_1 + 2x_2 + n_4 - p_4 = 380$$

The constraint for the manufacturing capacity is a hard constraint since the machine cannot produce more than its capacity. Hence, this constraint will not be transformed into a goal and it will remain as

$$x_1 + x_2 \leq 200$$

The constraints about the limit on the units of each product are hard constraints since the company will pay a high fee if it will not meet the demands of its customer. Hence, these constraints will not be transformed into goals and they will remain as

$$x_1 \geq 60$$

$$x_2 \geq 60$$

The aforementioned constraints are the explicit constraints. The explicit constraints are those that are explicitly given in the problem statement. This problem also has other constraints called implicit constraints. These are constraints that are not explicitly given in the problem statement but are present nonetheless. These constraints are typically associated with natural restrictions on the decision variables. In this problem, it is clear that one cannot have negative values for the amount of shirts and jackets that are produced. That is, x_1 and x_2 must be non-negative integer variables. Similarly, all deviational variables must be non-negative integer variables.

The entire Goal Programming problem can be formally stated as

$$\begin{array}{lrcrcrcrcl}
\min z = & p_1 & + & p_2 & + & n_3 & + & p_4 & & \\
\text{s.t.} & 2x_1 & + & 4x_2 & + & n_1 & - & p_1 & = & 600 \\
 & 5x_1 & + & 3x_2 & + & n_2 & - & p_2 & = & 700 \\
 & 100x_1 & + & 90x_2 & + & n_3 & - & p_3 & = & 18{,}000 \\
 & 2x_1 & + & 2x_2 & + & n_4 & - & p_4 & = & 380 \\
 & x_1 & + & x_2 & & & & & \leq & 200 \\
 & x_1 & & & & & & & \geq & 60 \\
 & & & x_2 & & & & & \geq & 60
\end{array}$$

$$x_1 \geq 0, x_2 \geq 0, \{x_1, x_2\} \in \mathbb{Z}$$
$$n_1 \geq 0, n_2 \geq 0, n_3 \geq 0, n_4 \geq 0, \{n_1, n_2, n_3, n_4\} \in \mathbb{Z}$$
$$p_1 \geq 0, p_2 \geq 0, p_3 \geq 0, p_4 \geq 0, \{p_1, p_2, p_3, p_4\} \in \mathbb{Z}$$

The careful reader will notice that in the objective function the unit measurements of the deviational variables are different; some are measured in m^2, one in € and another in hours. This is a common mistake in formulating a goal programming model as the units in the objective function need to be the same. In order to point this out, the authors chose this problem, as they have encountered often this issue in the literature. Otherwise the rationale of the goal programming principles presented in this Section is correct. There are many ways to remedy this issue: (i) use the weighted lexicographical goal programming variant where a normalization constant can be used to assure a commensurability of the goals (see Section 6.3), (ii) use the lexicographic goal programming variant where the goals can be prioritized (see Section 6.4), and (iii) use the Chebyshev goal programming variant that seeks to minimize the maximum unwanted deviation instead of the sum of deviations (see Section 6.5). Therefore, we are using this example in classical goal programming to point out that all variables should have the same scales and explore ways to deal with this situation in the following sections.

This goal programming problem is graphically represented in Figure 6.4. There are three hard constraints that imply the feasible region (again without taking into account that the variables are integers). All the other goals may be

- satisfied: if the optimal solution lies in a constraint's line, or
- under-achieved: if the optimal solution is in the direction of the associated negative deviational variable n, or
- over-achieved: if the optimal solution is in the direction of the associated positive deviational variable p.

Both the objective function and the constraints are linear. Hence, this is an integer linear programming problem. Solving this problem with an integer linear programming solver, like CPLEX, we derive the solution $x_1 = 81$ and $x_2 = 110$.

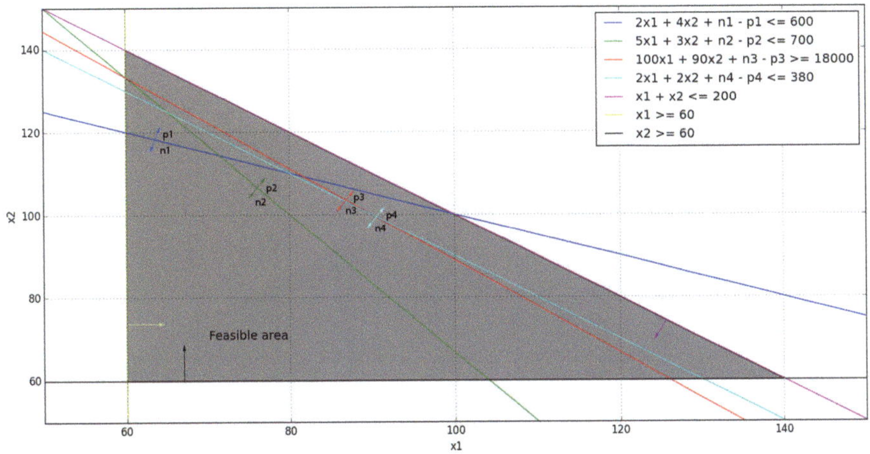

Fig. 6.4 Graphical representation of the Goal Programming problem

Table 6.2 Solution of the example with the classical Goal Programming method

Goal	Target value	Achieved value	Deviation (%)
Cotton	600	602	0.33
Linen	700	735	5
Profit	18,000	18,000	0
Production	380	382	0.53
Average	–	–	1.47

Table 6.2 and Figure 6.5 present information about the solution that was found. Only the goal for the profit was fully satisfied. All other constraints were over-achieved. The maximum deviation is found on the goal for the linen, where the company needs to buy additional 35 m^2 linen. The average deviation from the goals is 1.47%.

6.2.2 *Python Implementation*

The file *classicalGP.py* includes a Python implementation that shows how easy it is to use Pyomo to solve a Goal Programming problem using the classical Goal Programming method. Comments embedded in the code listing describe each part of the code. Initially, an object to perform optimization is created (line 9). Any integer programming solver can be used instead of CPLEX. Next, an object of a concrete model is created (line 12). Then, we define the decision variables (lines 15–16) and the deviational variables (lines 19–26) as non-negative integer variables. Next, we define the objective function (lines 29–30) and the constraints (lines 33–44) of the problem. Finally, we solve the Goal Programming problem (line 47) and print the values of the decision variables (lines 50–51) and the deviations from the target level of each goal (lines 54–88).

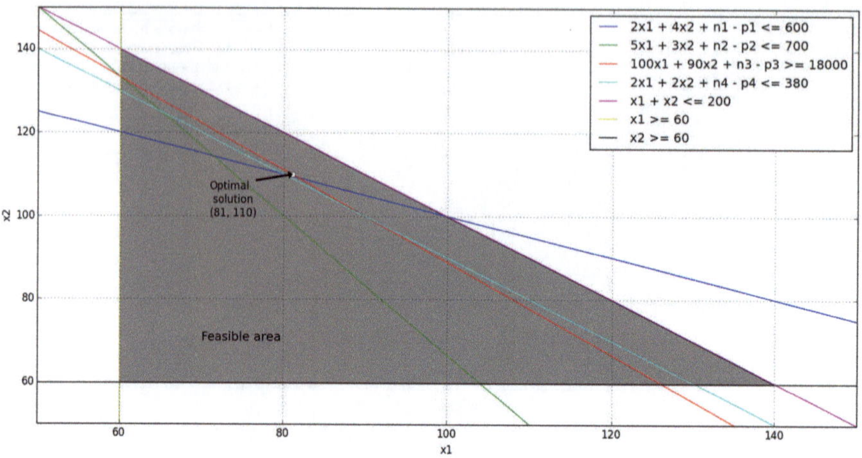

Fig. 6.5 Optimal solution of the example with the classical Goal Programming method

```
# Filename: classicalGP.py
# Description: Classical Goal Programming method
# Authors: Papathanasiou, J. & Ploskas, N.

from pyomo.environ import *
from pyomo.opt import SolverFactory

# Create an object to perform optimization
opt = SolverFactory('cplex')

# Create an object of a concrete model
model = ConcreteModel()

# Define the decision variables
model.x1 = Var(within = NonNegativeIntegers)
model.x2 = Var(within = NonNegativeIntegers)

# Define the deviational variables
model.n1 = Var(within = NonNegativeIntegers)
model.p1 = Var(within = NonNegativeIntegers)
model.n2 = Var(within = NonNegativeIntegers)
model.p2 = Var(within = NonNegativeIntegers)
model.n3 = Var(within = NonNegativeIntegers)
model.p3 = Var(within = NonNegativeIntegers)
model.n4 = Var(within = NonNegativeIntegers)
model.p4 = Var(within = NonNegativeIntegers)

# Define the objective function
model.obj = Objective(expr = model.p1 + model.p2 +
    model.n3 + model.p4)

# Define the constraints
```

```
model.con1 = Constraint(expr = 2 * model.x1 +
    4 * model.x2 + model.n1 - model.p1 == 600)
model.con2 = Constraint(expr = 5 * model.x1 +
    3 * model.x2 + model.n2 - model.p2 == 700)
model.con3 = Constraint(expr = 100 * model.x1 +
    90 * model.x2 + model.n3 - model.p3 == 18000)
model.con4 = Constraint(expr = 2 * model.x1 +
    2 * model.x2 + model.n4 - model.p4 == 380)
model.con5 = Constraint(expr = model.x1 +
    model.x2 <= 200)
model.con6 = Constraint(expr = model.x1 >= 60)
model.con7 = Constraint(expr = model.x2 >= 60)

# Solve the Goal Programming problem
opt.solve(model)

# Print the values of the decision variables
print("x1 = ", model.x1.value)
print("x2 = ", model.x2.value)

# Print the achieved values for each goal
if model.n1.value > 0:
    print("The first goal is underachieved by ",
          model.n1.value)
elif model.p1.value > 0:
    print("The first goal is overachieved by ",
          model.p1.value)
else:
    print("The first goal is fully satisfied")

if model.n2.value > 0:
    print("The second goal is underachieved by ",
          model.n2.value)
elif model.p2.value > 0:
    print("The second goal is overachieved by ",
          model.p2.value)
else:
    print("The second goal is fully satisfied")

if model.n3.value > 0:
    print("The third goal is underachieved by ",
          model.n3.value)
elif model.p3.value > 0:
    print("The third goal is overachieved by ",
          model.p3.value)
else:
    print("The third goal is fully satisfied")

if model.n4.value > 0:
    print("The fourth goal is underachieved by ",
          model.n4.value)
elif model.p4.value > 0:
    print("The fourth goal is overachieved by ",
          model.p4.value)
```

```
else:
    print("The fourth goal is fully satisfied")
```

The output of the code in this case (the solution) is as follows:

```
x1 = 81.0
x2 = 110.0
The first goal is overachieved by 2.0
The second goal is overachieved by 35.0
The third goal is fully satisfied
The fourth goal is overachieved by 2.0
```

6.3 Weighted Goal Programming

The first variant of a Goal Programming method presented in this section is called weighted Goal Programming. It can also be found in the literature as non-preemptive Goal Programming. The main idea of this variant is to attach penalty weights to the unwanted deviational variables in the objective function. These weights consist of two parts:

- The importance of the penalization at each deviational variable. We denote by u_i the weight associated with the minimization of n_i, while v_i is the weight associated with the minimization of p_i, where $i = 1, 2, \cdots, m$. These weights show the relative importance of the minimization of a deviational variable. Note that deviational variables whose minimization is unimportant, e.g., the negative deviational variable of a goal for cost, are assigned a weight equal to 0.
- A normalization constant, k_i, in order to assure a commensurability of the goals. These factors are necessary in order to scale all goals into the same units of measurement.

A weighted Goal Programming problem can be represented by the following formulation

$$
\begin{aligned}
\min z &= \sum_{i=1}^{m} \left(\frac{u_i n_i}{k_i} + \frac{v_i p_i}{k_i} \right) \\
\text{s.t.} \quad & f_i(x) + n_i + p_i = b_i \\
& x \in F \\
& n_i, p_i \geq 0, i = 1, 2, \ldots, m
\end{aligned}
$$

The above Goal Programming model is similar to the classical Goal Programming problem. The only difference is the objective function. Therefore, most of the steps that we follow to create a weighted Goal Programming problem are the same as described in Section 6.2. However, one additional step is needed; the selection of the weights by the decision maker. The numerator of the weight is chosen by the decision maker to depict the importance of the penalization at each deviational variable. As already mentioned, deviational variables whose minimization is unimportant, e.g., the negative deviational variable of a goal for cost, are assigned a weight equal to 0. Next, a normalization constant should be selected. There are many normalization techniques that have been used in weighted Goal Programming problems. The most widely-used are the following

1. Percentage normalization: In this normalization method, the denominator, k_i, is equal to the right-hand side value of the associated goal. Each deviational variable is turned into a percentage value away from its target value. Hence, all deviations are measured in the same units. The percentage normalization is simple and straightforward. However, it can cause some distortion when a subset of the goals is measured in the same units [12].
2. Zero-One normalization: In this normalization method, the deviational variables are scaled into a zero-one range. The value zero represents a deviation of zero, while the value one represents the worst possible deviation. The latter value is the denominator used when applying this normalization method. In order to find this value, we solve a single objective maximization linear programming problem for each goal. The objective value will be the worst possible deviation for that goal. If such a linear programming problem is unbounded, then the decision maker must choose a realistic estimation of the denominator. The drawbacks of this method are the following: (1) it requires the solution of m linear programming problems to find the worst deviation for each goal, and (2) in examples with unbounded goals, it can suffer from the problem of irrelevant values chosen by the decision maker.
3. Euclidean normalization: In this normalization method, the denominator, k_i, is equal to the Euclidean mean of the technical coefficients of the associated goal. For example, the denominator k_1 of the first goal in the example presented in the previous section, $2x_1 + 4x_2 + n_1 - p_1 = 600$, will be equal to $\sqrt{2^2 + 4^2} = \sqrt{20}$. The drawbacks of this method are the following [18]: (1) this method does not take into account the right-hand side values of the goals and this can lead to relatively low values of the normalization factors, and (2) when this method is used to normalize the deviations, the optimal value of the objective function does not have any obvious meaning.

To sum up, a weighted Goal Programming problem is formulated using the following steps

1. Determine whether a constraint is soft or hard.
2. Add a negative and a positive deviational variable on each constraint. Determine the type of the constraint and add to the objective function the deviational variable(s) to be penalized. For each deviational variable, select a weight.
3. Use a normalization method to scale the deviations.
4. Each hard constraint is written as a typical linear programming constraint.
5. Add bound constraints to the problem (if applicable).

6.3.1 Numerical Example

Following the steps described in the previous section, we will end up to the same constraints found when applying the classical Goal Programming method. In addition, we have to select the weights for the deviational variables to be minimized. In this example, we consider 1% deviation from the target of 380 man-hours is three times more important than the goals related to the cotton and linen. In addition, we consider 1% deviation from the target of 18,000 € profit is twice more important than the goals related to the cotton and linen. Therefore, the weights that we selected are the following:

- $u_1 = 0$ and $v_1 = 1$
- $u_2 = 0$ and $v_2 = 1$
- $u_3 = 2$ and $v_3 = 0$
- $u_4 = 0$ and $v_4 = 3$

Moreover, we will use the percentage normalization method. Hence, the denominator, k_i, is equal to the right-hand side value of the associated goal.

The weighted Goal Programming problem can be formally stated as

$$
\begin{array}{llcccccccl}
\min z = & \multicolumn{9}{l}{(1/600)p_1 + (1/700)p_2 + (2/18{,}000)n_3 + (3/380)p_4} \\
\text{s.t.} & 2x_1 & + & 4x_2 & + & n_1 & - & p_1 & = & 600 \\
 & 5x_1 & + & 3x_2 & + & n_2 & - & p_2 & = & 700 \\
 & 100x_1 & + & 90x_2 & + & n_3 & - & p_3 & = & 18{,}000 \\
 & 2x_1 & + & 2x_2 & + & n_4 & - & p_4 & = & 380 \\
 & x_1 & + & x_2 & & & & & \leq & 200 \\
 & x_1 & & & & & & & \geq & 60 \\
 & & & x_2 & & & & & \geq & 60
\end{array}
$$

$$x_1 \geq 0, x_2 \geq 0, \{x_1, x_2\} \in \mathbb{Z}$$
$$n_1 \geq 0, n_2 \geq 0, n_3 \geq 0, n_4 \geq 0, \{n_1, n_2, n_3, n_4\} \in \mathbb{Z}$$
$$p_1 \geq 0, p_2 \geq 0, p_3 \geq 0, p_4 \geq 0, \{p_1, p_2, p_3, p_4\} \in \mathbb{Z}$$

Solving this problem with an integer linear programming solver, like CPLEX, we derive the solution $x_1 = 80$ and $x_2 = 110$. Table 6.3 and Figure 6.6 present information about the solution that was found. The goals for the cotton and the production were fully satisfied. The constraint for the linen was over-achieved, while the constraint for the profit was under-achieved. The maximum deviation is found on the goal for the linen, where the company needs to buy additional 30 m^2 linen. The average deviation from the goals is 1.21%. This solution is better than the one found in the previous section when applying the classical Goal Programming method since the maximum and the average deviation is smaller.

Table 6.3 Solution of the example with the weighted Goal Programming method

Goal	Target value	Achieved value	Deviation (%)
Cotton	600	600	0
Linen	700	730	4.29
Profit	18,000	17,900	0.56
Production	380	380	0
Average	–	–	1.21

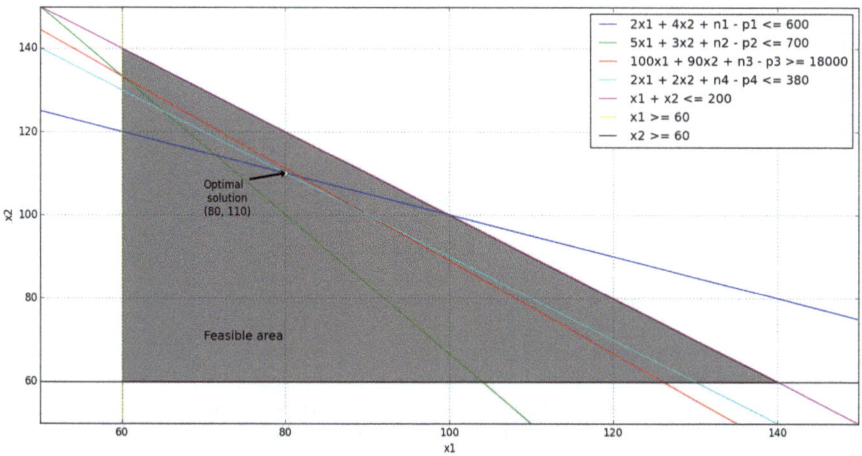

Fig. 6.6 Optimal solution of the example with the weighted Goal Programming method

6.3.2 Python Implementation

The file *weightedGP.py* includes a Python implementation that shows how easy it is to use Pyomo to solve a Goal Programming problem using the weighted Goal Programming method. Comments embedded in the code listing describe each part of the code. Initially, an object to perform optimization is created (line 9). Any integer programming solver can be used instead of CPLEX. Next, an object of a concrete model is created (line 12). Then, we define the decision variables (lines 15–16) and the deviational variables (lines 19–26) as non-negative integer variables. Next, we define the objective function (lines 30–32) and the constraints (lines 35–46) of the problem. Finally, we solve the Goal Programming problem (line 49) and print the values of the decision variables (lines 52–53) and the deviations from the target level of each goal (lines 56–90).

```
1.    # Filename: weightedGP.py
2.    # Description: Weighted Goal Programming method
3.    # Authors: Papathanasiou, J. & Ploskas, N.
4.
5.    from pyomo.environ import *
6.    from pyomo.opt import SolverFactory
```

```

# Create an object to perform optimization
opt = SolverFactory('cplex')

# Create an object of a concrete model
model = ConcreteModel()

# Define the decision variables
model.x1 = Var(within = NonNegativeIntegers)
model.x2 = Var(within = NonNegativeIntegers)

# Define the deviational variables
model.n1 = Var(within = NonNegativeIntegers)
model.p1 = Var(within = NonNegativeIntegers)
model.n2 = Var(within = NonNegativeIntegers)
model.p2 = Var(within = NonNegativeIntegers)
model.n3 = Var(within = NonNegativeIntegers)
model.p3 = Var(within = NonNegativeIntegers)
model.n4 = Var(within = NonNegativeIntegers)
model.p4 = Var(within = NonNegativeIntegers)

# Define the objective function with the
# associated weights (percentage normalization)
model.obj = Objective(expr = (1 / 600) * model.p1 +
    (1 / 700) * model.p2 + (2 / 18000) * model.n3 +
    (3 / 380) * model.p4)

# Define the constraints
model.con1 = Constraint(expr = 2 * model.x1 +
    4 * model.x2 + model.n1 - model.p1 == 600)
model.con2 = Constraint(expr = 5 * model.x1 +
    3 * model.x2 + model.n2 - model.p2 == 700)
model.con3 = Constraint(expr = 100 * model.x1 +
    90 * model.x2 + model.n3 - model.p3 == 18000)
model.con4 = Constraint(expr = 2 * model.x1 +
    2 * model.x2 + model.n4 - model.p4 == 380)
model.con5 = Constraint(expr = model.x1 +
    model.x2 <= 200)
model.con6 = Constraint(expr = model.x1 >= 60)
model.con7 = Constraint(expr = model.x2 >= 60)

# Solve the Goal Programming problem
opt.solve(model)

# Print the values of the decision variables
print("x1 = ", model.x1.value)
print("x2 = ", model.x2.value)

# Print the achieved values for each goal
if model.n1.value > 0:
    print("The first goal is underachieved by ",
          model.n1.value)
elif model.p1.value > 0:
    print("The first goal is overachieved by ",
```

```
               model.p1.value)
else:
    print("The first goal is fully satisfied")

if model.n2.value > 0:
    print("The second goal is underachieved by ",
          model.n2.value)
elif model.p2.value > 0:
    print("The second goal is overachieved by ",
          model.p2.value)
else:
    print("The second goal is fully satisfied")

if model.n3.value > 0:
    print("The third goal is underachieved by ",
          model.n3.value)
elif model.p3.value > 0:
    print("The third goal is overachieved by ",
          model.p3.value)
else:
    print("The third goal is fully satisfied")

if model.n4.value > 0:
    print("The fourth goal is underachieved by ",
          model.n4.value)
elif model.p4.value > 0:
    print("The fourth goal is overachieved by ",
          model.p4.value)
else:
    print("The fourth goal is fully satisfied")
```

The output of the code in this case (the solution) is as follows

```
x1 = 80.0
x2 = 110.0
The first goal is fully satisfied
The second goal is overachieved by 30.0
The third goal is underachieved by 100.0
The fourth goal is fully satisfied
```

6.4 Lexicographic Goal Programming

The second variant of a Goal Programming method presented in this section is called lexicographic Goal Programming. It can also be found in the literature as preemptive Goal Programming. The main feature of this variant is the existence of a number of priority levels. This variant is used when the decision maker has a clear preference order for satisfying the goals. Each priority level consists of a number of unwanted deviations to be minimized. We define by L the number of priority levels with corresponding index $l = 1, 2, \cdots, L$. Each priority level is a function of a subset of unwanted deviational variables, $h_l(n, p)$. The consensus in the goal programming literature is that no more than five priority levels should be used in this variant [12].

A lexicographic Goal Programming problem can be represented by the following formulation

$$
\begin{aligned}
\min z = & \; h_1(n, p), h_2(n, p), \cdots, h_L(n, p) \\
\text{s.t.} \quad & f_i(x) + n_i + p_i = b_i \\
& x \in F \\
& n_i, p_i \geq 0, i = 1, 2, \ldots, m
\end{aligned}
$$

The above Goal Programming model is similar to the classical Goal Programming problem. The only difference is the objective function. Therefore, most of the steps that we follow to create a lexicographic Goal Programming problem are the same as described in Section 6.2. However, there is one major difference. We need to solve L linear programming problems. Each linear programming problem has in its objective the subset of the unwanted deviational variables that we want to minimize in each step. In addition, we also add constraints to each linear programming problem concerning the optimal value of the unwanted variables minimized in the previous priority levels. Therefore, the linear programming problem of the second priority level will have a constraint about the optimal value of the unwanted deviational variables minimized in the linear programming problem of the first priority level; the linear programming problem of the third priority level will have two constraints about the optimal values of the unwanted deviational variables minimized in the linear programming problems of the first and the second priority level, etc.

To sum up, a lexicographic Goal Programming problem is formulated using the following steps

1. Determine whether a constraint is soft or hard.
2. Add a negative and a positive deviational variable on each constraint. Determine the type of the constraint and add to the objective function the deviational variable(s) to be penalized.
3. Each hard constraint is written as a typical linear programming constraint.
4. Add bound constraints to the problem (if applicable).
5. Formulate and solve L linear programming problems. The objective of the lth, where $l = 1, 2, \cdots, L$, linear programming problem includes a subset of the unwanted deviational variables that will be minimized in this priority level. If $l > 1$, add constraint(s) to the linear programming problem concerning the optimal value of the unwanted variable(s) mimimized in the previous priority level(s).

6.4.1 Numerical Example

Following the steps described in the previous section, we will end up to the same constraints found when applying the classical Goal Programming method. In addition, we have to determine an order of preference for the goals. In this example, we consider that the company wishes to see the goals satisfied in the following order:

1. Achieve the production goal.
2. Achieve the profit goal.
3. Achieve the goal for the cotton and the linen.

Hence, the objective function of each priority level will be the following:

1. $h_1(p_4)$
2. $h_2(n_3)$
3. $h_3(p_1, p_2)$

The Goal Programming problem of the first priority level can be formally stated as

$$
\begin{array}{lrrrrcl}
\min z = & p_4 & & & & & \\
\text{s.t.} & 2x_1 & + 4x_2 & + n_1 & - p_1 & = & 600 \\
 & 5x_1 & + 3x_2 & + n_2 & - p_2 & = & 700 \\
 & 100x_1 & + 90x_2 & + n_3 & - p_3 & = & 18{,}000 \\
 & 2x_1 & + 2x_2 & + n_4 & - p_4 & = & 380 \\
 & x_1 & + x_2 & & & \leq & 200 \\
 & x_1 & & & & \geq & 60 \\
 & & x_2 & & & \geq & 60
\end{array}
$$
$$
\begin{array}{l}
x_1 \geq 0, x_2 \geq 0, \{x_1, x_2\} \in \mathbb{Z} \\
n_1 \geq 0, n_2 \geq 0, n_3 \geq 0, n_4 \geq 0, \{n_1, n_2, n_3, n_4\} \in \mathbb{Z} \\
p_1 \geq 0, p_2 \geq 0, p_3 \geq 0, p_4 \geq 0, \{p_1, p_2, p_3, p_4\} \in \mathbb{Z}
\end{array}
$$

Solving this problem with an integer linear programming solver, like CPLEX, we derive the solution $p_4 = 0$. Hence, we can formulate the Goal Programming problem of the second priority level. We should also add the constraint $p_4 = 0$ to the new problem.

$$
\begin{array}{lrrrrcl}
\min z = & n_3 & & & & & \\
\text{s.t.} & 2x_1 & + 4x_2 & + n_1 & - p_1 & = & 600 \\
 & 5x_1 & + 3x_2 & + n_2 & - p_2 & = & 700 \\
 & 100x_1 & + 90x_2 & + n_3 & - p_3 & = & 18{,}000 \\
 & 2x_1 & + 2x_2 & + n_4 & - p_4 & = & 380 \\
 & x_1 & + x_2 & & & \leq & 200 \\
 & x_1 & & & & \geq & 60 \\
 & & x_2 & & & \geq & 60 \\
 & p_4 & & & & = & 0
\end{array}
$$
$$
\begin{array}{l}
x_1 \geq 0, x_2 \geq 0, \{x_1, x_2\} \in \mathbb{Z} \\
n_1 \geq 0, n_2 \geq 0, n_3 \geq 0, n_4 \geq 0, \{n_1, n_2, n_3, n_4\} \in \mathbb{Z} \\
p_1 \geq 0, p_2 \geq 0, p_3 \geq 0, p_4 \geq 0, \{p_1, p_2, p_3, p_4\} \in \mathbb{Z}
\end{array}
$$

Table 6.4 Solution of the example with the lexicographic Goal Programming method

Goal	Target value	Achieved value	Deviation (%)
Cotton	600	580	3.33
Linen	700	750	7.14
Profit	18,000	18,000	0
Production	380	380	0
Average	–	–	2.62

Solving this problem with an integer linear programming solver, like CPLEX, we derive the solution $n_3 = 0$. Hence, we can formulate the Goal Programming problem of the third priority level. We should also add the constraint $n_3 = 0$ to the new problem. Note that the price per m^2 for both cotton and linen can be considered the same remedying both materials fully interchangeable and comparable. This way we can safely assume that the unit measurements of the deviational variables p_1 and p_2 are the same.

$$\begin{array}{llcl}
\min z = & p_1 + p_2 & & \\
\text{s.t.} & 2x_1 + 4x_2 + n_1 - p_1 & = & 600 \\
& 5x_1 + 3x_2 + n_2 - p_2 & = & 700 \\
& 100x_1 + 90x_2 + n_3 - p_3 & = & 18{,}000 \\
& 2x_1 + 2x_2 + n_4 - p_4 & = & 380 \\
& x_1 + x_2 & \leq & 200 \\
& x_1 & \geq & 60 \\
& x_2 & \geq & 60 \\
& p_4 & = & 0 \\
& n_3 & = & 0 \\
\multicolumn{4}{l}{x_1 \geq 0, x_2 \geq 0, \{x_1, x_2\} \in \mathbb{Z}} \\
\multicolumn{4}{l}{n_1 \geq 0, n_2 \geq 0, n_3 \geq 0, n_4 \geq 0, \{n_1, n_2, n_3, n_4\} \in \mathbb{Z}} \\
\multicolumn{4}{l}{p_1 \geq 0, p_2 \geq 0, p_3 \geq 0, p_4 \geq 0, \{p_1, p_2, p_3, p_4\} \in \mathbb{Z}}
\end{array}$$

Solving this problem with an integer linear programming solver, like CPLEX, we derive the solution $x_1 = 90$ and $x_2 = 100$. Table 6.4 and Figure 6.7 present information about the solution that was found. The goals for the profit and the production were fully satisfied. The constraint for the cotton was under-achieved, while the constraint for the linen was over-achieved. The maximum deviation is found on the goal for the linen, where the company needs to buy additional 50 m^2 linen. The average deviation from the goals is 2.62%. This solution is worse than the one found in the previous section when applying the weighted Goal Programming method since the maximum and the average deviation is larger. However, it satisfies the first two goals in the preference order that the company set; the goals for the production and profit are fully satisfied.

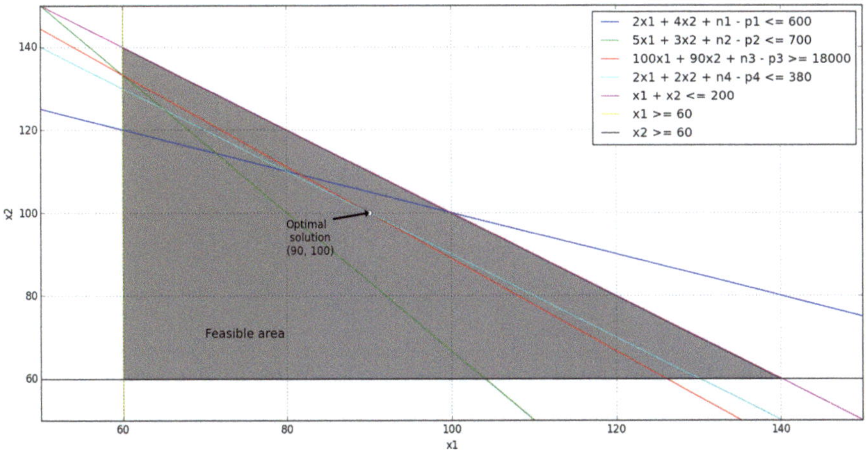

Fig. 6.7 Optimal solution of the example with the lexicographic Goal Programming method

6.4.2 Python Implementation

The file *lexicographicGP.py* includes a Python implementation that shows how easy it is to use Pyomo to solve a Goal Programming problem using the lexicographic Goal Programming method. Comments embedded in the code listing describe each part of the code. Initially, an object to perform optimization is created (line 10). Any integer programming solver can be used instead of CPLEX. Next, an object of a concrete model is created (line 13). Then, we define the decision variables (lines 16–17) and the deviational variables (lines 20–27) as non-negative integer variables. Next, we define the objective function of the first priority level (line 31) and the constraints (lines 34–45) of the problem. We solve the Goal Programming problem of the first priority level (line 49) and retrieve the value of the first priority level (line 52). Then, we define the objective function of the second priority level (line 56) and add a constraint for the value of the first priority level (line 60). We solve the Goal Programming problem of the second priority level (line 64) and retrieve the value of the second priority level (line 67). Next, we define the objective function of the third priority level (line 71) and add a constraint for the value of the second priority level (line 75). Finally, we solve the Goal Programming problem (line 79) and print the values of the decision variables (lines 82–83) and the deviations from the target level of each goal (lines 86–120).

```
# Filename: lexicographicGP.py
# Description: Lexicographic Goal Programming
# method
# Authors: Papathanasiou, J. & Ploskas, N.

from pyomo.environ import *
from pyomo.opt import SolverFactory

# Create an object to perform optimization
opt = SolverFactory('cplex')

# Create an object of a concrete model
model = ConcreteModel()

# Define the decision variables
model.x1 = Var(within = NonNegativeIntegers)
model.x2 = Var(within = NonNegativeIntegers)

# Define the deviational variables
model.n1 = Var(within = NonNegativeIntegers)
model.p1 = Var(within = NonNegativeIntegers)
model.n2 = Var(within = NonNegativeIntegers)
model.p2 = Var(within = NonNegativeIntegers)
model.n3 = Var(within = NonNegativeIntegers)
model.p3 = Var(within = NonNegativeIntegers)
model.n4 = Var(within = NonNegativeIntegers)
model.p4 = Var(within = NonNegativeIntegers)

# Define the objective function of the
# first priority level
model.obj = Objective(expr = model.p4)

# Define the constraints
model.con1 = Constraint(expr = 2 * model.x1 +
    4 * model.x2 + model.n1 - model.p1 == 600)
model.con2 = Constraint(expr = 5 * model.x1 +
    3 * model.x2 + model.n2 - model.p2 == 700)
model.con3 = Constraint(expr = 100 * model.x1 +
    90 * model.x2 + model.n3 - model.p3 == 18000)
model.con4 = Constraint(expr = 2 * model.x1 +
    2 * model.x2 + model.n4 - model.p4 == 380)
model.con5 = Constraint(expr = model.x1 +
    model.x2 <= 200)
model.con6 = Constraint(expr = model.x1 >= 60)
model.con7 = Constraint(expr = model.x2 >= 60)

# Solve the Goal Programming problem of the
# first priority level
opt.solve(model)

# Retrieve the value of the first priority level
p4 = model.p4.value

# Define the objective function of the
```

```
# second priority level
model.obj = Objective(expr = model.n3)

# Add a constraint for the value of the first
# priority level
model.con8 = Constraint(expr = model.p4 == p4)

# Solve the Goal Programming problem of the
# second priority level
opt.solve(model)

# Retrieve the value of the second priority level
n3 = model.n3.value

# Define the objective function of the
# third priority level
model.obj = Objective(expr = model.p1 + model.p2)

# Add a constraint for the value of the second
# priority level
model.con9 = Constraint(expr = model.n3 == n3)

# Solve the Goal Programming problem of the
# third priority level
opt.solve(model)

# Print the values of the decision variables
print("x1 = ", model.x1.value)
print("x2 = ", model.x2.value)

# Print the achieved values for each goal
if model.n1.value > 0:
    print("The first goal is underachieved by ",
          model.n1.value)
elif model.p1.value > 0:
    print("The first goal is overachieved by ",
          model.p1.value)
else:
    print("The first goal is fully satisfied")

if model.n2.value > 0:
    print("The second goal is underachieved by ",
          model.n2.value)
elif model.p2.value > 0:
    print("The second goal is overachieved by ",
          model.p2.value)
else:
    print("The second goal is fully satisfied")

if model.n3.value > 0:
    print("The third goal is underachieved by ",
          model.n3.value)
elif model.p3.value > 0:
    print("The third goal is overachieved by ",
```

```
            model.p3.value)
else:
    print("The third goal is fully satisfied")

if model.n4.value > 0:
    print("The fourth goal is underachieved by ",
            model.n4.value)
elif model.p4.value > 0:
    print("The fourth goal is overachieved by ",
            model.p4.value)
else:
    print("The fourth goal is fully satisfied")
```

The output of the code in this case (the solution) is as follows

```
x1 = 90.0
x2 = 100.0
The first goal is underachieved by 20.0
The second goal is overachieved by 50.0
The third goal is fully satisfied
The fourth goal is fully satisfied
```

6.5 Chebyshev Goal Programming

The third variant of a Goal Programming method presented in this section is called Chebyshev Goal Programming. This variant was introduced by Flavell [6] and it is known as Chebyshev Goal Programming, because it uses the Chebyshev distance or $L\infty$ metric. It can also be found in the literature as Minmax Goal Programming. The main idea of this variant is to achieve a balance between the goals. Classical, weighted and lexicographic Goal Programming often find extreme solutions, i.e., points that lie in the intersection of goals, constraints, and axes. This can lead to an unbalanced solution since some goals are achieved and others are far from satisfactory. In Chebyshev Goal Programming, we introduce additional constraints in order to ensure balance between the goals. This is the only widely-used variant that can find optimal solutions that are not located at extreme points.

Let λ be the maximal deviation from amongst the set of goals, then a generic form of a Chebyshev Goal Programming problem is the following

$$
\begin{aligned}
\min z = \quad & \lambda \\
\text{s.t.} \quad & \tfrac{u_i n_i}{k_i} + \tfrac{v_i p_i}{k_i} + \quad \leq \lambda \\
& f_i(x) + n_i + p_i = b_i \\
& x \in F \\
n_i, p_i \geq 0, i = 1, 2, \ldots, m
\end{aligned}
$$

The above Goal Programming model is similar to the classical Goal Programming problem. The only difference is the objective function and m new constraints that limit the unwanted deviational variables to be equal or less than λ. Therefore, most of the steps that we follow to create a Chebyshev Goal Programming problem are the same as described in Section 6.2.

To sum up, a Chebyshev Goal Programming problem is formulated using the following steps:

1. Determine whether a constraint is soft or hard.
2. Add a negative and a positive deviational variable on each constraint. Determine the type of the constraint and add a constraint of the deviational variable(s) to be penalized. For each deviational variable, select a weight.
3. Use a normalization method to scale the deviations.
4. Each hard constraint is written as a typical linear programming constraint.
5. Add bound constraints to the problem (if applicable).

6.5.1 Numerical Example

Following the steps described in Section 6.2, we will end up to the same constraints found when applying the classical Goal Programming method. In addition, we have to select the weights for the deviational variables to be minimized. In this example, we consider 1% deviation from the target of 380 man-hours is three times more important than the goals related to the cotton and linen. In addition, we consider 1% deviation from the target of 18,000 € profit is twice more important than the goals related to the cotton and linen. Therefore, the weights that we selected are the following:

- $u_1 = 0$ and $v_1 = 1$
- $u_2 = 0$ and $v_2 = 1$
- $u_3 = 2$ and $v_3 = 0$
- $u_4 = 0$ and $v_4 = 3$

Moreover, we will use the percentage normalization method. Hence, the denominator, k_i, is equal to the right-hand side value of the associated goal.

Table 6.5 Solution of the example with the Chebyshev Goal Programming method

Goal	Target value	Achieved value	Deviation (%)
Cotton	600	614	2.33
Linen	700	716	2.29
Profit	18,000	17,830	0.94
Production	380	380	0
Average	–	–	1.39

The Chebyshev Goal Programming problem can be formally stated as

$$
\begin{array}{llcl}
\min z = & \lambda & & \\
\text{s.t.} & (1/600)p_1 & \leq & \lambda \\
 & (1/700)p_2 & \leq & \lambda \\
 & (2/18{,}000)n_3 & \leq & \lambda \\
 & (3/380)p_4 & \leq & \lambda \\
 & 2x_1 + 4x_2 + n_1 - p_1 & = & 600 \\
 & 5x_1 + 3x_2 + n_2 - p_2 & = & 700 \\
 & 100x_1 + 90x_2 + n_3 - p_3 & = & 18{,}000 \\
 & 2x_1 + 2x_2 + n_4 - p_4 & = & 380 \\
 & x_1 + x_2 & \leq & 200 \\
 & x_1 & \geq & 60 \\
 & x_2 & \geq & 60
\end{array}
$$

$$\lambda \geq 0$$
$$x_1 \geq 0, x_2 \geq 0, \{x_1, x_2\} \in \mathbb{Z}$$
$$n_1 \geq 0, n_2 \geq 0, n_3 \geq 0, n_4 \geq 0, \{n_1, n_2, n_3, n_4\} \in \mathbb{Z}$$
$$p_1 \geq 0, p_2 \geq 0, p_3 \geq 0, p_4 \geq 0, \{p_1, p_2, p_3, p_4\} \in \mathbb{Z}$$

Solving this problem with an integer linear programming solver, like CPLEX, we derive the solution $x_1 = 73$ and $x_2 = 117$. Table 6.5 and Figure 6.8 present information about the solution that was found. Only the goal for the production was fully satisfied. The goals for the cotton and linen were over-achieved, while the goal for the profit was under-achieved. The maximum deviation is found on the goal for the linen, where the company needs to buy additional $16\,\text{m}^2$ linen. The average deviation from the goals is 1.39%. Note that the maximum deviation of this solution is smaller than the maximum deviation found in the previous examples of this chapter when applying the classical, the weighted, and the lexicographic Goal Programming methods. On the other hand, while weighted and lexicographic Goal Programming result in solutions that two goals are fully satisfied, Chebyshev Goal Programming results in a solution where only one goal is fully satisfied.

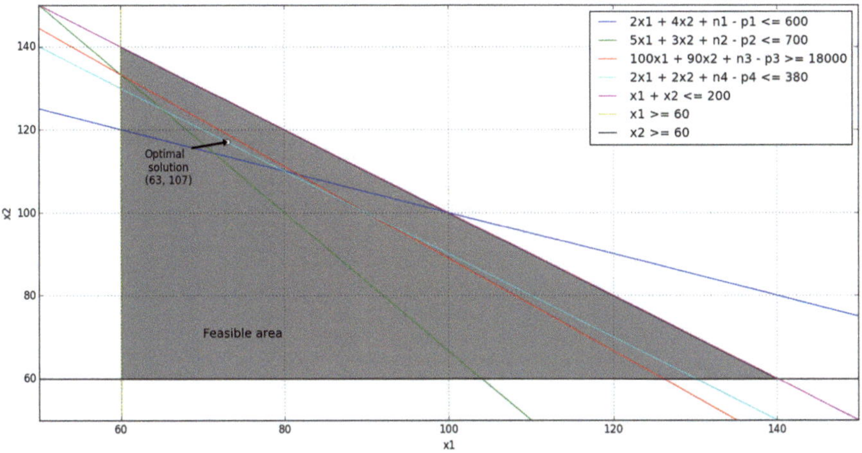

Fig. 6.8 Optimal solution of the example with the Chebyshev Goal Programming method

6.5.2 *Python Implementation*

The file *chebyshevGP.py* includes a Python implementation that shows how easy it is to use Pyomo to solve a Goal Programming problem using the Chebyshev Goal Programming method. Comments embedded in the code listing describe each part of the code. Initially, an object to perform optimization is created (line 9). Any integer programming solver can be used instead of CPLEX. Next, an object of a concrete model is created (line 12). Then, we define the decision variables (lines 15–16) and the deviational variables (lines 19–26) as non-negative integer variables. We also define the variable of maximal deviation from amongst the set of goals (line 30). Next, we define the objective function (line 33) and the constraints (lines 36–55) of the problem. Finally, we solve the Goal Programming problem (line 58) and print the values of the decision variables (lines 61–62) and the deviations from the target level of each goal (lines 65–99).

```
1.   # Filename: chebyshevGP.py
2.   # Description: Chebyshev Goal Programming method
3.   # Authors: Papathanasiou, J. & Ploskas, N.
4.
5.   from pyomo.environ import *
6.   from pyomo.opt import SolverFactory
7.
8.   # Create an object to perform optimization
9.   opt = SolverFactory('cplex')
10.
11.  # Create an object of a concrete model
12.  model = ConcreteModel()
13.
```

```
# Define the decision variables
model.x1 = Var(within = NonNegativeIntegers)
model.x2 = Var(within = NonNegativeIntegers)

# Define the deviational variables
model.n1 = Var(within = NonNegativeIntegers)
model.p1 = Var(within = NonNegativeIntegers)
model.n2 = Var(within = NonNegativeIntegers)
model.p2 = Var(within = NonNegativeIntegers)
model.n3 = Var(within = NonNegativeIntegers)
model.p3 = Var(within = NonNegativeIntegers)
model.n4 = Var(within = NonNegativeIntegers)
model.p4 = Var(within = NonNegativeIntegers)

# Define the variable of maximal deviation
# from amongst the set of goals
model.l = Var(within=NonNegativeReals)

# Define the objective function
model.obj = Objective(expr = model.l)

# Define the constraints
model.con1 = Constraint(expr = (1 / 600) *
    model.p1 <= model.l)
model.con2 = Constraint(expr = (1 / 700) *
    model.p2 <= model.l)
model.con3 = Constraint(expr = (2 / 18000) *
    model.n3 <= model.l)
model.con4 = Constraint(expr = (3 / 380) *
    model.p4 <= model.l)
model.con5 = Constraint(expr = 2 * model.x1 +
    4 * model.x2 + model.n1 - model.p1 == 600)
model.con6 = Constraint(expr = 5 * model.x1 +
    3 * model.x2 + model.n2 - model.p2 == 700)
model.con7 = Constraint(expr = 100 * model.x1 +
    90 * model.x2 + model.n3 - model.p3 == 18000)
model.con8 = Constraint(expr = 2 * model.x1 +
    2 * model.x2 + model.n4 - model.p4 == 380)
model.con9 = Constraint(expr = model.x1 +
    model.x2 <= 200)
model.con10 = Constraint(expr = model.x1 >= 60)
model.con11 = Constraint(expr = model.x2 >= 60)

# Solve the Goal Programming problem
opt.solve(model)

# Print the values of the decision variables
print("x1 = ", model.x1.value)
print("x2 = ", model.x2.value)

# Print the achieved values for each goal
if model.n1.value > 0:
    print("The first goal is underachieved by ",
          model.n1.value)
```

```
elif model.p1.value > 0:
    print("The first goal is overachieved by ",
          model.p1.value)
else:
    print("The first goal is fully satisfied")

if model.n2.value > 0:
    print("The second goal is underachieved by ",
          model.n2.value)
elif model.p2.value > 0:
    print("The second goal is overachieved by ",
          model.p2.value)
else:
    print("The second goal is fully satisfied")

if model.n3.value > 0:
    print("The third goal is underachieved by ",
          model.n3.value)
elif model.p3.value > 0:
    print("The third goal is overachieved by ",
          model.p3.value)
else:
    print("The third goal is fully satisfied")

if model.n4.value > 0:
    print("The fourth goal is underachieved by ",
          model.n4.value)
elif model.p4.value > 0:
    print("The fourth goal is overachieved by ",
          model.p4.value)
else:
    print("The fourth goal is fully satisfied")
```

The output of the code in this case (the solution) is as follows

```
x1 = 73.0
x2 = 117.0
The first goal is overachieved by 14.0
The second goal is overachieved by 16.0
The third goal is underachieved by 170.0
The fourth goal is fully satisfied
```

References

1. Brown, K., & Norgaard, R. (1992). Modelling the telecommunication pricing decision. *Decision Sciences, 23*(3), 673–686.
2. Charnes, A., Cooper, W. W. (1981). *Management models and industrial applications of linear programming*. Hoboken: Wiley.
3. Charnes, A., Cooper, W. W., & Ferguson, R. O. (1955). Optimal estimation of executive compensation by linear programming. *Management Science, 1*(2), 138–151.

4. Charnes, A., Cooper, W. W., Harrald, J., Karwan, K. R., & Wallace, W. A. (1976). A goal interval programming model for resource allocation in a marine environmental protection program. *Journal of Environmental Economics and Management, 3*(4), 347–362.
5. Charnes, A., Duffuaa, S., & Al-Saffar, A. (1989). A dynamic goal programming model for planning food self-sufficiency in the Middle East. *Applied Mathematical Modelling, 13*(2), 86–93.
6. Flavell, R. B. (1976). A new goal programming formulation. *Omega, 4*(6), 731–732.
7. Hart, W. E., Laird, C., Watson, J. P., & Woodruff, D. L. (2012). *Pyomo-optimization modeling in python* (Vol. 67). Berlin: Springer.
8. Hoffman, J. J., & Schniederjans, M. J. (1990). An international strategic management/goal programming model for structuring global expansion decisions in the hospitality industry: The case of Eastern Europe. *International Journal of Hospitality Management, 9*(3), 175–190.
9. Ignizio, J. P. (1976). *Goal programming and extensions*. New York: Lexington Books.
10. Ignizio, J. P. (1981). Antenna array beam pattern synthesis via goal programming. *European Journal of Operational Research, 6*(3), 286–290.
11. Ijiri, Y. (1965). *Management goals and accounting for control* (Vol. 3). Amsterdam: North Holland.
12. Jones, D., & Tamiz, M. (2010). *Practical goal programming* (Vol. 141). New York: Springer.
13. Khorramshahgol, R., & Azani, H. (1988). A decision support system for effective systems analysis and planning. *Journal of Information and Optimization Sciences, 9*(1), 41–52.
14. Lee, S. M. (1972). *Goal programming for decision analysis*. Philadelphia: Auerbach.
15. Piech, B., & Rehman, T. (1993). Application of multiple criteria decision making methods to farm planning: A case study. *Agricultural Systems, 41*(3), 305–319.
16. Puelz, A. V., & Lee, S. M. (1992). A multiple-objective programming technique for structuring tax-exempt serial revenue debt issues. *Management Science, 38*(8), 1186–1200.
17. Pyomo (2016). http://www.pyomo.org/ (Current as of 30 April, 2017).
18. Romero, C. (1991). *A handbook of critical issues in goal programming*. Oxford: Pergamon Press.
19. Schniederjans, M. (1995). *Goal programming: Methodology and applications*. Berlin: Springer.
20. Schniederjans, M. J., & Hoffman, J. (1992). Multinational acquisition analysis: A zero-one goal programming model. *European Journal of Operational Research, 62*(2), 175–185.
21. Tamiz, M., & Jones, D. F. (1996). Goal programming and Pareto efficiency. *Journal of Information and Optimization Sciences, 17*(2), 291–307. ISO 690.

Appendix: Revised Simos

A.1 Introduction

An important and challenging step in the solution of a decision making problem is the elicitation of weights. In the examples presented in the previous Chapters, the weights of the criteria were inputs to the problem. However, one of the most important steps that the decision maker performs during the solution of a decision making problem with an MCDA method is the assessment of the criteria weights. Various methods have been proposed for this task. These methods can be mainly categorized in two major classes, the direct assessment ones and the indirect ones. The indirect assessment methods have been used in most applications of MCDA methods due to their simplicity and realism. One of the most widely-used methods is the one proposed by Simos [2, 3]. This method is a typical indirect assessment method that has been widely-used in decision making problems since it is relatively easy for decision makers to express their preferences. The elicitation of the criteria weights is performed by asking decision makers to express the relative importance of the criteria, through the arrangement of criteria cards, from the least to the most important one. The method was later extended by Figuera and Roy [1] in order to address certain robustness issues of the original method. The revised Simos method is widely-used in decision making problems for estimating the criteria weights.

A.2 Methodology

The decision maker is given a set of cards. The name of each criterion is written on a card. The decision maker uses the cards in order to rank the criteria from the least important to the most important. The first criterion in the ranking is the least important and the last criterion is the most important. If some criteria have the same importance for the decision maker, he/she can place them together in the same

J. Papathanasiou, N. Ploskas, *Multiple Criteria Decision Aid*, Springer Optimization and Its Applications 136, https://doi.org/10.1007/978-3-319-91648-4

position. Therefore, a complete pre-order of the whole n criteria is obtained. The number of ranks is $\widetilde{n}$, where $1 \leq \widetilde{n} \leq n$ (since some of the cards can be placed in the same rank).

The decision maker has also a set of white cards. The importance of two successive criteria (or two successive subsets of ex aequo criteria in case two or more cards have been placed together) in the ranking can be more or less close. In order to depict this smaller or larger difference of the importance of successive criteria, the decision maker introduces white cards between two successive cards. The more the number of white cards between two successive criteria, the greater the difference between their importance. If no white card is placed between two successive ranks, then the difference between the weights of the criteria in these two successive ranks can be chosen as the unit, u, for measuring the intervals between weights. Hence, if one white card is placed between two successive ranks, then there is a difference of $2u$ between the weights of the criteria in these two successive ranks. Finally, the decision maker should state how many times the last criterion is more important than the first one. This ratio is denoted by the parameter z.

The revised Simos method consists of two steps:

1. Calculate the non-normalized weights $k = (k_1, k_2, \cdots, k_{\widetilde{n}})$: Let $e_r{}'$ be the number of white cards between the ranks r and $r + 1$.

$$\begin{cases} e_r = e_r{}' + 1, \forall r = 1, 2, \cdots, \widetilde{n} - 1, e_1{}' = 0 \\ u = \frac{z-1}{\sum_{r=1}^{\widetilde{n}-1} e_i} \end{cases} \tag{A.1}$$

Next, we can calculate the non-normalized weights k as follows:

$$k_r = 1 + u \sum_{i=0}^{r-1} e_i, e_0 = 0 \tag{A.2}$$

2. Calculate the normalized weights $k^* = \left(k_1^*, k_2^*, \cdots, k_{\widetilde{n}}^*\right)$: Let c_i be the number of cards in each ranking i, where $1 \leq i \leq \widetilde{n}$. Then, the normalized weights k^* are calculated as follows:

$$k_r^* = \frac{100}{\sum_{i=1}^{\widetilde{n}} c_i k_i} k_r \tag{A.3}$$

Figuera and Roy [1] also presented a method to eliminate some of the decimal figures from the normalized weights.

A.2.1 Numerical Example

Let us consider a set of eight criteria: $\{a, b, c, d, e, f, g, h\}$. Let us also suppose that the decision maker groups together cards associated to the criteria having the same importance into four subsets of ex aequo, inserting also three white cards between some of the successive ranks. Table A.1 presents the ranking of these cards.

Let us suppose that $z = 6.5$, i.e., the decision maker states that the last subset is 6.5 times more important than the first one. We start by calculating vector e and parameter u according to Equation (A.1)

$$e = \begin{bmatrix} 1 & 2 & 3 & 1 \end{bmatrix}$$

$$u = 1.375$$

Then, we can calculate the non-normalized weights k according to Equation (A.2)

$$k = \begin{bmatrix} 1 & 2.375 & 5.125 & 9.25 \end{bmatrix}$$

Finally, we calculate the normalized weights k^* according to Equation (A.3)

$$k^* = \begin{bmatrix} 2.61437908 & 6.20915033 & 13.39869281 & 24.18300654 \end{bmatrix}$$

Therefore, the criteria weights are presented in Table A.2.

Table A.1 Ranking of the criteria using cards

Rank	Subset of ex aequo
1	$\{b, d\}$
2	$\{c\}$
3	White card
4	$\{e, f, h\}$
5	White card
6	White card
7	$\{a, g\}$

Table A.2 Criteria weights

Criterion	Weight
a	24.1830065359
b	2.61437908497
c	6.2091503268
d	2.61437908497
e	13.3986928105
f	13.3986928105
g	24.1830065359
h	13.3986928105
Total	100

A.2.2 Python Implementation

The file *RevisedSimos.py* includes a Python implementation of the revised Simos method. The input variables are array *subsets* (the ranks of the criteria, lines 8–9), and z (the parameter z, line 12). In lines 15–22, the number of cards, the number of positions with non-white cards, and vector c are calculated. Then, we calculate parameter u (line 25). Next, we calculate vector e (lines 29–34). The non-normalized weights are computed in lines 37–41, while the normalized weights are computed in lines 44–46. Finally, the criteria weights are printed in lines 49–57.

```
1.  # Filename: RevisedSimos.py
2.  # Description: Revised Simos method
3.  # Authors: Papathanasiou, J. & Ploskas, N.
4.
5.  from numpy import *
6.
7.  # placement of cards ('w' represents a white card)
8.  subsets = array([['b','d'], ['c'], ['w'],
9.      ['e', 'f', 'h'], ['w'], ['w'], ['a', 'g']])
10.
11. # parameter z
12. z = 6.5
13.
14. # calculate number of cards, positions, and vector c
15. noOfcards = 0
16. positions = 0
17. c = []
18. for i in range(subsets.shape[0]):
19.     if subsets[i][0] != 'w':
20.         noOfcards = noOfcards + len(subsets[i][:])
21.         positions = positions + 1
22.         c.append(len(subsets[i][:]))
23.
24. # calculate u
25. U = round((z - 1) / positions, 6)
26.
```

```
# calculate vector e
e = ones(positions)
counter = -1
for i in range(subsets.shape[0]):
    if subsets[i][0] != 'w':
        counter = counter + 1
    else:
        e[counter] = e[counter] + 1

# calculate the non-normalized weights k
k = ones(positions)
totalk = k[0] * c[0]
for i in range(1, positions):
    k[i] = 1 + U * sum(e[0:i])
    totalk = totalk + k[i] * c[i]

# calculate the normalized weights
normalizedWeights = zeros(positions)
for i in range(0, positions):
    normalizedWeights[i] = (100 / totalk) * k[i]

# print the criteria weights
counter = -1
for i in range(subsets.shape[0]):
    if subsets[i][0] != 'w':
        counter = counter + 1
    else:
        continue
    for j in range(len(subsets[i][:])):
        print("Weight of criterion ", subsets[i][j],
           " = ", normalizedWeights[counter])
```

References

1. Figueira, J., & Roy, B. (2002). Determining the weights of criteria in the ELECTRE type methods with a revised Simos' procedure. *European Journal of Operational Research, 139*(2), 317–326.
2. Simos, J. (1990). *Evaluer l'impact sur l'environnement: Une approche originale par l'analyse multicritère et la négociation*. Lausanne: Presses polytechniques et universitaires romandes.
3. Simos, J. (1990). *L'évaluation environnementale: Un processus cognitif négocié*. Ph.D. dissertation, DGF-EPFL, Lausanne, France.

Index

J. Papathanasiou, N. Ploskas, *Multiple Criteria Decision Aid*, Springer Optimization and Its Applications 136, https://doi.org/10.1007/978-3-319-91648-4

MIX
Papier aus verantwortungsvollen Quellen
Paper from responsible sources
FSC® C105338

If you have any concerns about our products,
you can contact us on
ProductSafety@springernature.com

In case Publisher is established outside the EU,
the EU authorized representative is:
Springer Nature Customer Service Center GmbH
Europaplatz 3, 69115 Heidelberg, Germany

Printed by Libri Plureos GmbH
in Hamburg, Germany